Electromagnetic Interference Issues in Power Electronics and Power Systems

Editor

Firuz Zare

Queensland University of Technology
Australia

eBooks End User License Agreement

CONTENTS

FOREWORD

The purpose of this e-Book is to address advanced topics of Electromagnetic Interference (EMI) in modern power electronics and power systems. This book aims at practicing engineers and graduate students and will be of immense help to them while analyzing EMI problems in electrical and electronic systems.

This e-Book focuses on conducted and radiated emission noise generated by different power converters such as Switch Mode Power Supplies and DC-AC Converters. EMI filter design and different approaches to predict common mode and differential mode noise are discussed in detail. One of the challenging issues in power converters is the thermal problem due to heat generated by switching devices. In this e-book, the effects of heatsink on conducted and radiated noise have been taken into account. Common mode and surge voltage issues in AC machines have also been addressed and discussed in this e-Book.

This e-Book consists of five chapters which address different EMI issues in power electronics and power systems. The chapters of this e-Book are written by eminent researchers and engineers. I had a great pleasure reading their material and would like to congratulate all the authors for their excellent contributions.

Overall, this e-Book addresses the most important electromagnetic interferences issues in Power Electronics and Power Systems.

Prof. Arindam Ghosh

IEEE Fellow
Queensland University of Technology

PREFACE

The purpose of this e-Book is to address advanced topics of Electromagnetic Interference (EMI) in modern power electronics and power systems, which can help engineers and students to analyze EMI problems in electrical and electronic systems.

This e-Book focuses on conducted and radiated emission noise generated by different power converters such as Switch Mode power Supplies and DC-AC Inverters. EMI filter design and different approaches to predict common mode and differential mode noise are illustrated in detail. One of the challenging issues in power converters is thermal problem due to heat generated by switching devices; the effects of heatsink on conducted and radiated noise have been taken into account. Common mode and surge voltage issues in AC machines have also been addressed and discussed in this e-Book.

This e-Book consists of five chapters which address different EMI issues in power electronics and power systems. The chapters of this e-Book were written by eminent researchers and engineers and I would like to thank all the authors for their time and contributions.

Firuz Zare
Queensland University of Technology
Australia

List of Contributors

J.P.S. Catalão

Department of Electromechanical Engineering, University of Beira Interior, R. Fonte do Lameiro, 6201-001, Covilha, Portugal

F. Canavero

Dipartimento di Elettronica, Politecnico di Torino, Torino, Corso Duca degli, Abruzzi, 24 - 10129, Torino, Italy

G. K. Felic

National ICT Australia, Victoria Research Laboratory, The University of Melbourne, Parkville VIC 3010, Australia

F. Leferink

Thales Nederland, Hengelo, Haaksbergerstraat 49, 7554 PA Hengelo, the Netherlands

J. Łuszcz

Faculty of Electrical and Control Engineering, Gdańsk University of Technology, G. Narutowicza 11/12, 80-952 Gdańsk, Poland

V.M.F. Mendes

Department of Electrical Engineering and Automation, Instituto Superior de Engenharia de Lisboa, R. Conselheiro Emídio Navarro, 1950-062 Lisbon, Portugal

R.B. Rodrigues

Department of Electrical Engineering and Automation, Instituto Superior de Engenharia de Lisboa, R. Conselheiro Emídio Navarro, 1950-062 Lisbon, Portugal

A. Roc'h

Department of Telecommunication Engineering, University of Twente, Drienerlolaan 5, 7522, NB Enschede, the Netherlands

K. Y. See

School of Electrical and Electronic Engineering, Nanyang Technological University, Block S1, Level B1C, Room 100, Nanyang Link, Singapore, 639798

V. Tarateeraseth

College of Data Storage Innovation, King Mongkut's Institute of Technology, Ladkrabang, Chalongkrung Rd., Ladkrabang, Bangkok, Thailand, 10520

CHAPTER 1

Analysis of Common Mode Inductors and Optimization Aspects

Anne Roc'h[1,]* and Frank Leferink[1,2]

[1]*Department of Telecommunication Engineering, University of Twente, Drienerlolaan 5, 7522 NB Enschede, The Netherlands and* [2]*Thales Nederland, Hengelo, Haaksbergerstraat 49, 7554 PA Hengelo, The Netherlands.*

Abstract: The common mode inductors, or common mode chokes, are a key component of electromagnetic interference filters. Engineers usually face significant design challenges such as size, cost and weight while working with these components. To avoid the construction of several prototypes, often oversized, engineers need an analytical method to predict performances of the filter. The analysis of the common mode inductors starts with a presentation of the different ferromagnetic materials: their properties are the cornerstone of the design of the component. Based on this study, the impedances used to characterize the choke are related to the designable parameters. The derived model shows the role of the parasitic currents to ground and the common mode impedance in the overall common mode current attenuation. A certain attenuation of differential mode current is also to be expected due to the parasitic currents of the common mode choke itself: the turn to turn capacitance and the leakage inductance. This model is validated by measurements. Sensitivity studies provide an additional insight into the behavior of the choke by giving an understanding on how variations of parameters, influence the final performance. The deviation calculation and the influence of the designable parameters are addressed at the end of this chapter.

Keywords: Capacitive coupling, Common mode current, Core material, Curie temperature, Differential mode current, Electromagnetic interference (EMI), Equivalent circuit, Ferromagnetic material, Ferromagnetism, Flux, Ferrite, Filter, Hysteresis loop, Heat, Inductance, Impedance, Iron, Modification factors, Metal alloys, Noise source, Nanocrystalline, Noise spectrum, Powder material, Permeability, Power converter, Radiated electric field, Saturation level, Sensitivity.

INTRODUCTION

Power electronic converters consist of a switching devices which are turned on or off, based on a modulation technique to adjust output voltage and/or frequency. They can be classified as AC-to-DC, DC-to-DC, DC-to-AC and AC-to-AC [1]. Fig. **1** contains a general presentation of the structure of frequency converters. They have a main drawback: the high level of electromagnetic interferences (EMI) generated on power lines and motor cables, due to the non-ideal behaviour of the switches. The noise spectrum is usually spread from around 10 kHz to several decades of MHz. Filtering the main supply and motor cables adequately is a challenge which often leads to retro-designed filters, tested in a 'cut and try' process [2], [3], [4]. Design constraints as size, cost and weight are common while working with such components. Availability of analytical methods to predict performances of the filters would reduce or avoid the need for construction of several, often oversized, prototypes.

Common mode (CM) currents are frequently referred at 'antenna-mode' currents and are the predominant mechanism for producing radiated electric field in practical products [5]. Fig. **3** presents the circulation of common mode current in a motor drive. These currents usually flow along the cable and return to the source *via* stray capacitances between the cable and the ground and to certain extent *via* capacitive coupling between the frame of the motor and the ground.

Passive filtering of common mode currents involves a combination of capacitors and common mode chokes because the value of the capacitance to ground is limited. The common mode choke is a key element in terms of performances, size, cost and weight of the overall filter and is the focus of this chapter.

The two following examples show the impact of the common mode filter on the level of electromagnetic interferences: Fig. **1** presents the attenuation of current at the output of a common-mode filter used in an AC/DC converter in which a MnZn ferrite of 10 mH is used in combination with two feed through Y-capacitors. The impact

---

*****Address correspondence to Anne Roc'h:** Department of Telecommunication Engineering, University of Twente, Drienerlolaan 5, 7522 NB Enschede, The Netherlands; E-mail: anne.roch@gmail.com

of a common mode filter on the level of radiated emission between 10kHz and 1GHz is shown in Fig. **2**. This filter has been built with nanocrystalline cores and constitutes a good example of an alternative design to the classical iron choke solution which will be detailed in this chapter. Designs of both these filters are detailed in [3] and [6].

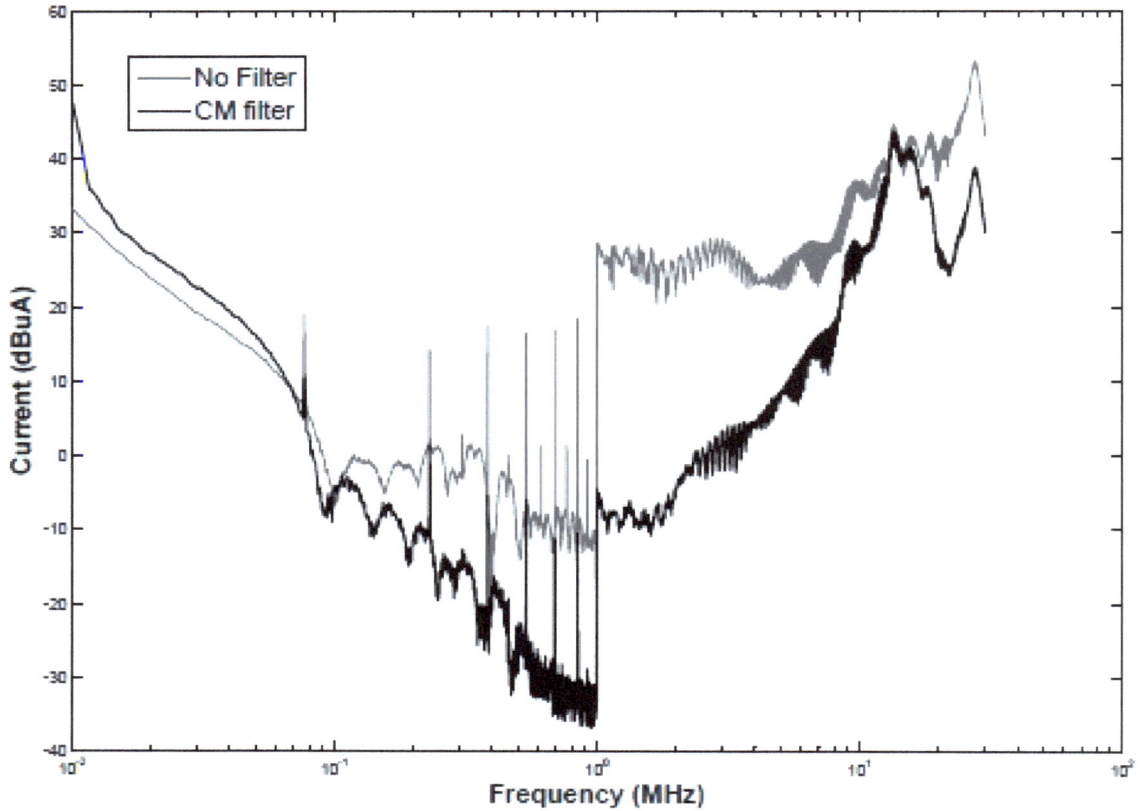

Figure 1: Input common mode current attenuation of an input filter of an AC/DC converter from 10kHz to 30MHz (level in dBµA).

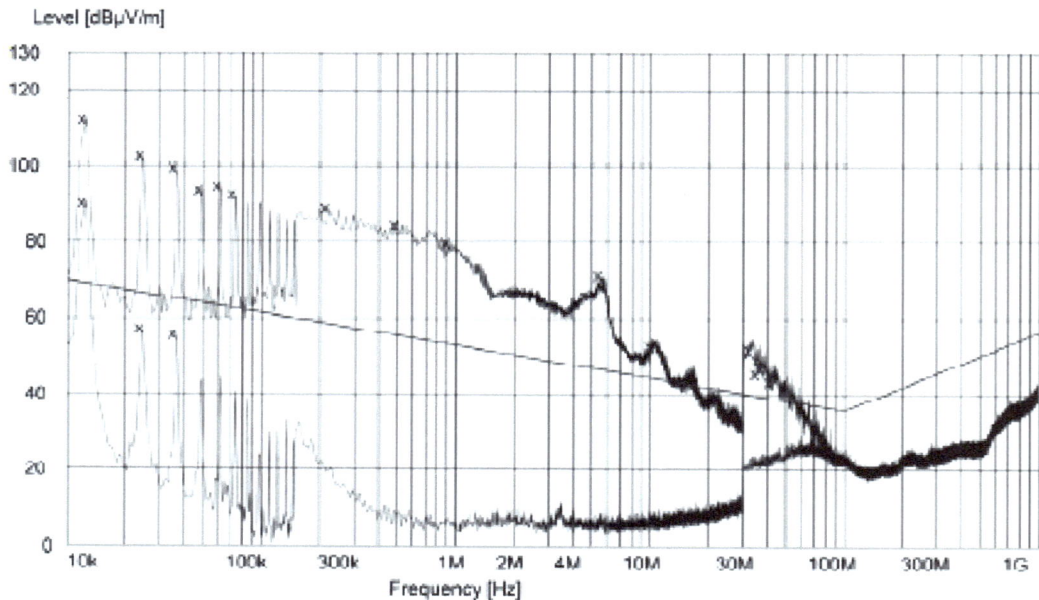

Figure 2: Radiated emission attenuation of an output filter for frequency converter from 10kHz to 1GHz (Level in dBµV/m).

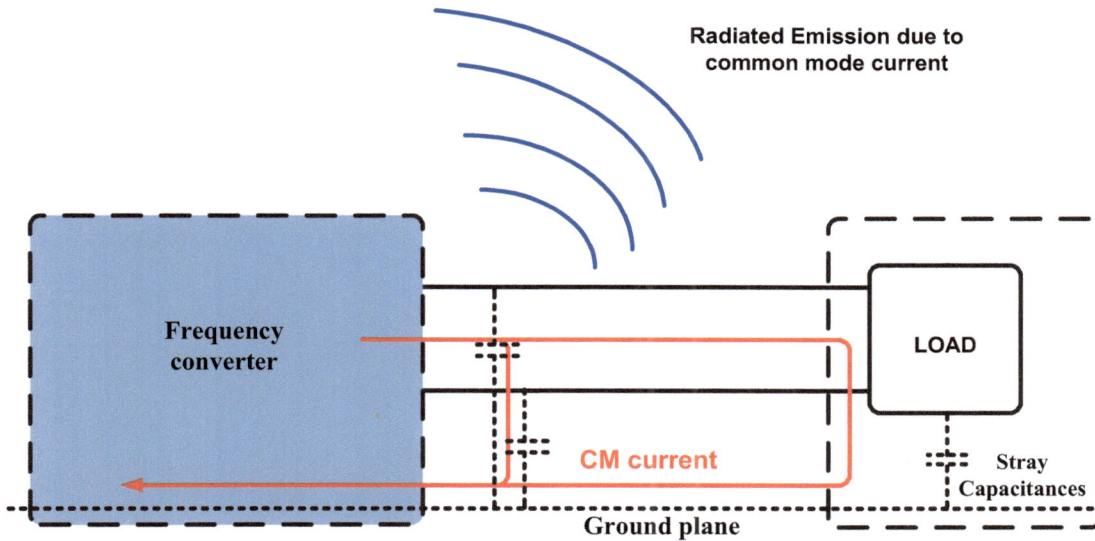

Figure 3: The common mode current path in a motor drive.

CORE MATERIALS

The core material can be used to increase the inductance, or to absorb energy and to transformed it in heat. The main purpose of a CM choke is not for energy storage but it is for a transformation of this energy to heat. The amount of common mode current that will be transform in heat by the common mode choke and the frequency range of efficiency will be first determined by the material itself and then by the overall design.

Ferromagnetic materials are typically used: iron powder material, ferrites, meta-materials and nanocrystalline. In this section the general material properties are first reviewed. The levels of permeability and saturation as well as the Curie temperature of materials are compared with each other.

Core Material Properties: Overview

Ferromagnetism. In a ferromagnetic material there is parallel alignment of the atomic moment in a domain. Each domain thus becomes a magnet. Their size and geometry are formed to reduce the magnetic potential contained in the field lines connecting north to south outside the material. Each domain contains about 10^{15} atoms. In this condition the magnetic flux path never leaves the boundary of the material. The region where the magnetization is the same, is called a 'domain wall'. When an electromagnetic flux is created across a ferromagnetic material the domains become aligned to produce a strong magnetic field within the part. Electromagnetic energy is transferred to the core which stored or transformed into heat depending on the frequency. The main purpose of a common mode choke though is not for energy storage but for the transformation of this energy into heat.

Permeability. In order to evaluate the total flux density in a ferromagnetic material it is useful to define the permeability μ which is the ratio of the magnetic flux density B, to magnetizing field H. It is the most important parameter used to characterize a magnetic material. The relative permeability μ_r is the ratio of the permeability μ *to* μ_o, where $\mu_o = 4.\pi * 10^{-7} \, H.m^{-1}$ is the permeability of free space.

The common mode choke is used for its core loss properties that are related to the imaginary part of the complex permeability and less for its inductive properties (related to the real part of the complex permeability).

Hysteresis loop and saturation level. Presentation. As the current goes through one sine-wave cycle, the magnetization goes through one hysteresis loop cycle. Minor hysteresis loops are obtained when the maximum applied field is lower than that required to saturate the material. Fig. **4** presents the shape of a hysteresis loop in a ferromagnetic material.

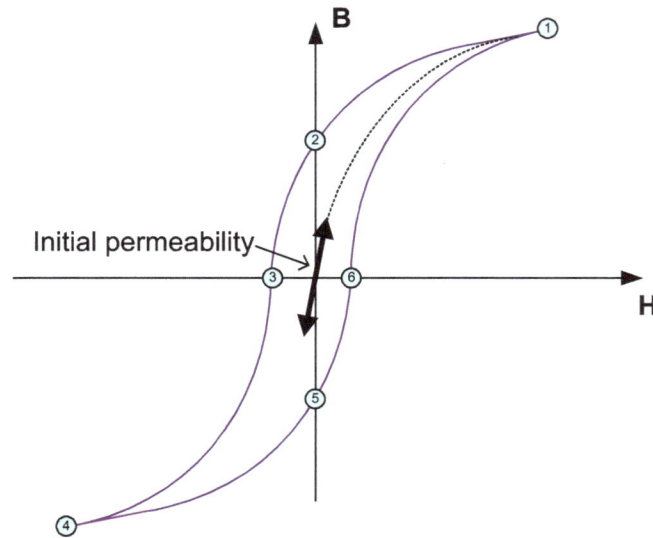

Figure 4: Hysteresis loop in a ferromagnetic material.

Hysteresis loop modeling. Permeability of the material is the slope of the *BH* loop and reduces close to the saturation. At this stage the common mode choke will not absorb energy anymore, and the common mode impedance is low. For these two reasons it is important for the designer to have knowledge about the shape of the *B-H*-loop.

Several approaches [13] have been developed to predict the major *B-H* loop of ferrites. In particular the Jiles-Atherton model [14] is convenient. It comprises of a first order non-linear differential equation which can be solved numerically to give the magnetization *M*, as a function of the applied magnetic field *H*.

The *B-H* loop is dependent of the topology of the cores and the main drawback of this method in a predictive model which is the need of experimental extraction of the so-called 'Jiles Atherton parameters' on the common mode choke itself. A solution is to combine the Jiles-Atherton model with the extraction parameters algorithm as described in [17].

Hysteresis loop in transient condition. The hysteresis loop as described in the previous section and its related Jiles-Atherton model are valid only when the current goes through a sine-wave cycle. In [15] the concept of volume fraction is introduced to model hysteresis loop of ferrite cores excited by a transient magnetic field. The Jiles-Atherton model is modified in a loop by loop technique in order to obtain the magnetization trajectory with an asymmetric minor loop excursion under arbitrary waveform magnetic field excitation. This technique is specially developed for power transformers and chokes used in power electronics converters. For high frequency operation (above 400 Hz) an additional technique has to be incorporated in this model to deal with rate of dependent hysteresis effect NL Mi, Oruganti R, SX Chen. Modeling of hysteresis loops of ferrite cores excited by a transient magnetic field Magnetics, IEEE Trans 1998; 34(4): 1294-1296

[16]. The hysteresis loop does indeed enlarge as the frequency increases and the permeability on the other hand decreases.

These models in transient conditions remain experimental: for each core and each wiring system tested on this core, five experimental parameters have to be initially extracted. This is due to the structure of the Jiles-Atherton model. They are useful for an accurate prediction of waveforms and hysteresis losses in a time domain.

Choke materials belong to the soft magnetic material family and are characterized by a narrower *B-H* loop. For low power application and/or in a choke used relatively far from its saturation condition, the permeability (slope coefficient of the hysteresis loop) remains similar to the initial permeability. The initial permeability is the parameter usually provided by manufacturer.

Curie temperature. The spontaneous magnetization of ferromagnetic material disappears at the Curie temperature. It is then important to ensure that the temperature of the choke remains below this temperature.

General Material Overview

The typical requirements of an optimized common mode choke are:

–   A high impedance over the wanted frequency-range, this is related to the complex permeability of the material,

–   A high saturation level,

Three basic materials: ferrites, powder materials (iron) and metal alloys (nanocrystalline and amorphous structure), are used in the design of the traditional common mode choke for switched mode power supplies (SMPS). Fig. **5** presents a synthetic overview of the permeability of nanocrystalline, ferrites, iron and amorphous materials [20-23].

The highest initial permeabilities are found in nanocrystalline materials (up to 10 000 till 150 kHz). Iron powder cores have low permeabilities (10 to 100 till 100 kHz) while amorphous alloys have higher values (5000 to 80 000 till 100 kHz). Ferrite cores can be used over a wide frequency range (up to several MHz for the NiZn cores). Higher flux density can be found in nanocrystalline and amorphous material, as well as in iron powder core. Ferrites exhibit a significantly lower level of saturation; this is depicted in Fig. **6**.

Iron based nanocrystalline materials are the new generation of magnetic alloys. Typical characteristic of these materials is their small nanometer grain size. Concept of nanocrystalline materials was introduced in 1981 [25]. Research on this new material has progressed after the development of the first nanocrystalline material in 1988.

Nanocrystalline materials are a good alternative to the traditional chokes made of ferrite or iron powder: the main advantage of this material is its high levels of saturation and in its relatively smaller size, it is however more expensive than the other materials. This newer material is a major step towards volume reduction, reaching 50-80% lower compared to a ferrite core and more than 90% lower compared to an iron core. The consequent weight loss can be an important property in aerospace applications or any other area where weight is a design constraint.

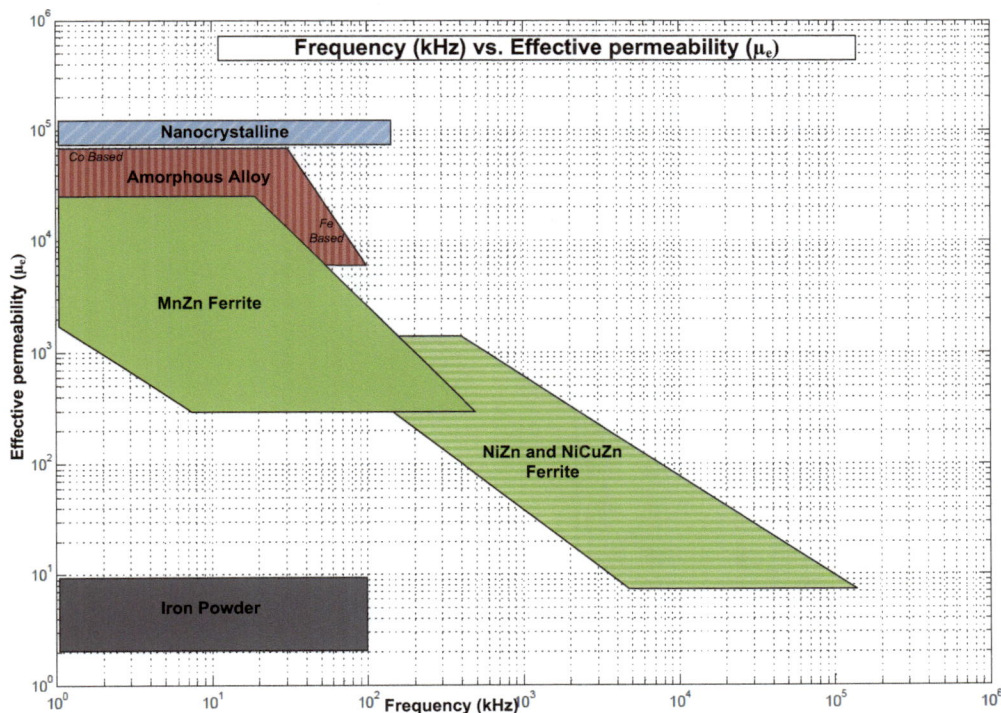

Figure 5: Main magnetic properties for the material ferrites, nanocrystalline and amorphous: Permeability vs. Frequency.

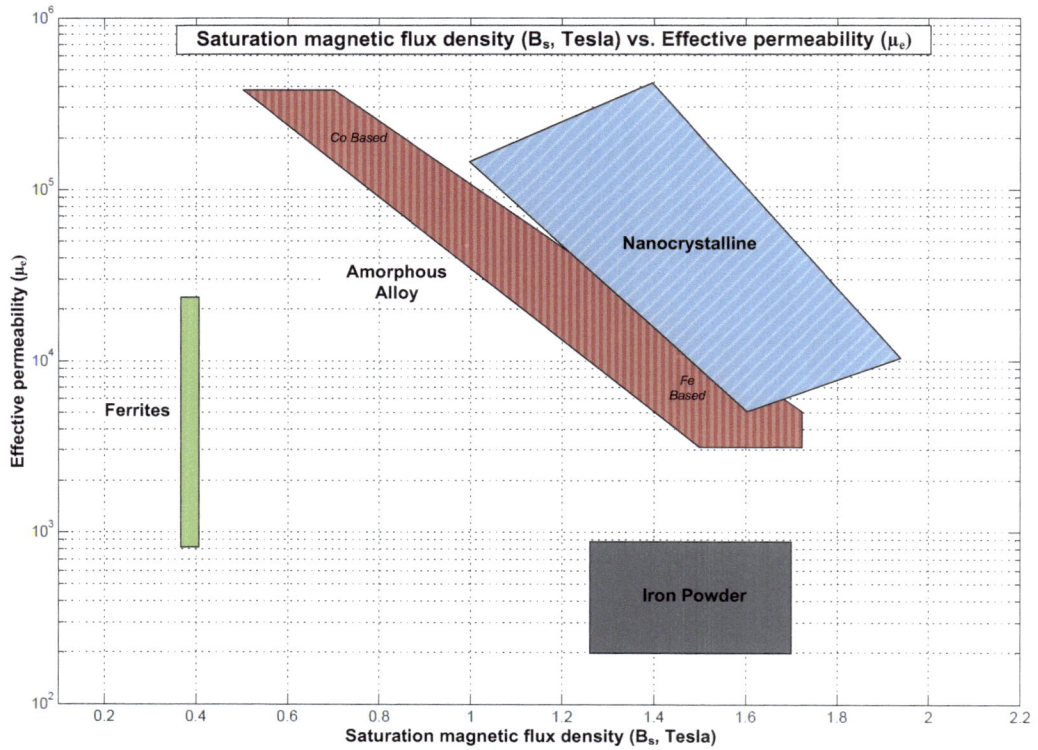

Figure 6: Main magnetic properties for the material ferrites, nanocrystalline and amorphous: Saturation vs. permeability.

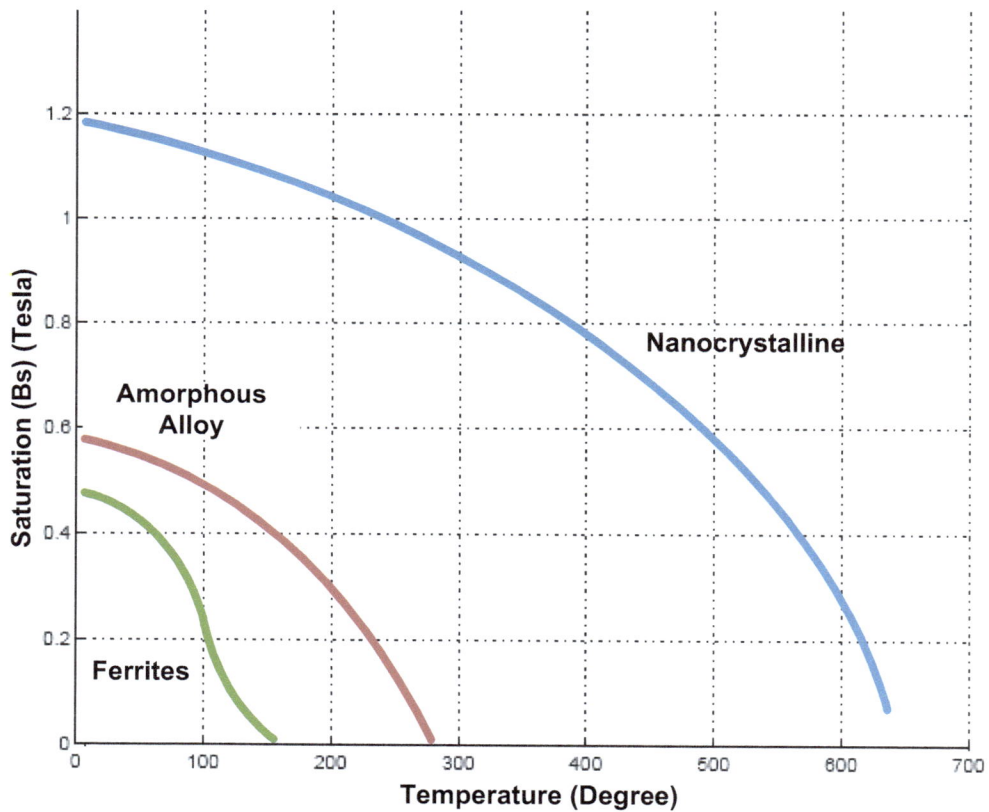

Figure 7: Temperature vs. Saturation.

Iron remains however the lowest cost option of the four family of materials. An equivalent ferrite solution would be about 4 times more expensive. A nanocrystalline or an amorphous solution would be between 7 and 10 times more costly than an iron solution [24]. Fig. **7** presents the evolution of the saturation level with the Curie temperature.

COMMON MODE CHOKES, IMPEDANCES AND DESIGNABLE PARAMETERS

This section focuses on the designable parameters of the common mode choke (CMC) and their related impedance modeling.

Common Mode Chokes Presentation

A main objective of a common mode choke is to block the emission noise (electromagnetic interferences) while the lower frequency range of the signals are not affected. Common mode inductors are wound with two or three windings of equal numbers of turns. The number of windings is same as the number of phases. As depicted in Fig. **8**, the windings are placed on the core so that the line currents in each winding create fluxes that are equal in magnitude but opposite in phase in the case of differential currents, and identical in the case of common mode currents. The fluxes of the differential mode currents are thus ideally cancelling out each other and the related current is not influenced by the inductors. It will be shown later that the cancellation of the fluxes cannot be achieved completely. The fluxes due to the common mode currents are in the same direction. These currents go together through equivalent common mode impedance related to the material properties and are transformed into heat.

Impedances and Designable Parameters Identification

The designable parameters are the quantitative aspects of physical characteristics of the common mode choke that are input to its design process. In this section the designable parameters of a common mode choke are related to their impedances in its equivalent circuit.

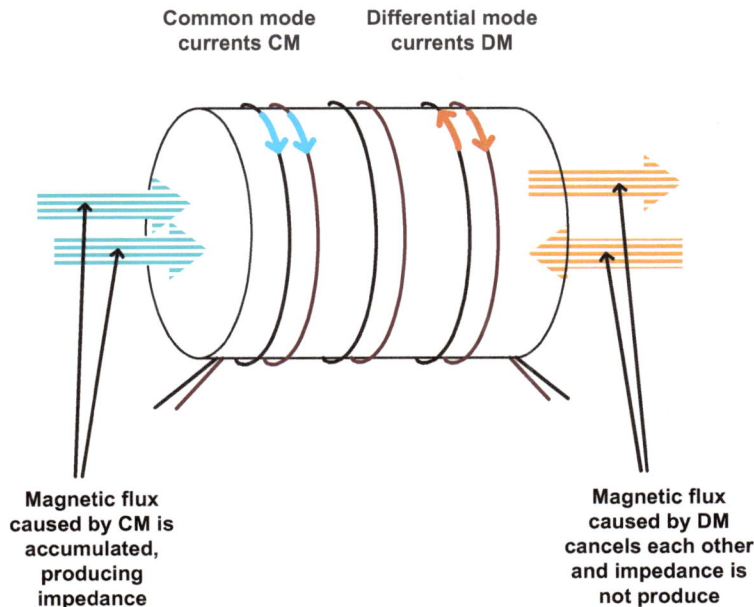

Figure 8: Magnetic flux in a common mode choke.

Fig. **9** presents a typical common mode choke under consideration: a choke with two symmetrical windings. The shape of the core is a toroidal. The study of the behavior of a common mode choke involves four types of impedances:

- *Common mode impedance*; it is the impedance faced by the common mode current in the choke.

- *Differential mode impedance*; it is the impedance faced by the differential mode current in the choke.

- *Inter-winding capacitance*; it is the parasitic impedance existing between two turns of a winding.

- *Intra winding capacitance*; it is the parasitic impedance existing between the two windings of the choke.

Table **1** lists the impedances of the common mode choke and their respective designable parameters. These designable parameters are the parameters that can be modified by the designer and will be the inputs of the behavioral model of the common mode choke.

Fig. **10** presents the general equivalent circuit of the common mode choke. The structure is validated *via* measurement in the next section. Z_{cm} is the common mode impedance, Z_{dm} is the differential mode impedance, C_{int} is the parasitic inter-winding capacitance and C_{tt} the parasitic intra-winding capacitance.

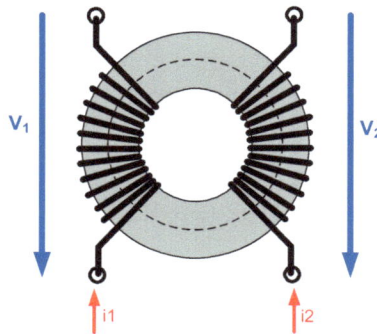

Figure 9: Typical CMC under consideration.

Table 1: Designable parameter for common mode choke

Impedances of the CMC	Designable parameters
Z_{cm}	Material (complex permeability) Dimension of the choke Number of turns Effective length
Z_{dm}	Number of turns Dimensions of the choke Angle of the winding free section
Inter and Intra Winding Capacitance	Number of turns Dimensions of the choke Wire dimensions and materials(isolation and diameter)

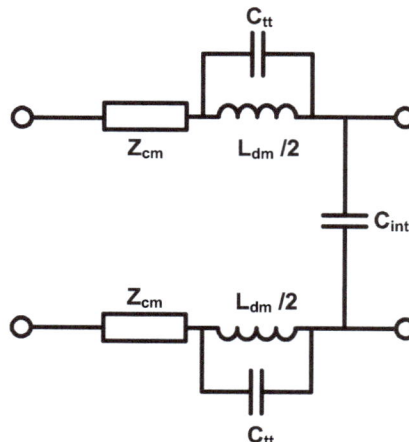

Figure 10: General equivalent circuit of the common mode choke.

Related Impedance Modeling

Common mode impedance. The value of the CM impedance is strongly related to the value of the permeability of the core. As detailed in [10] the inductor introduces frequency variable impedances in the circuit. Permeability of a common mode inductor is a complex parameter, the real component represents the reactive component and the imaginary part represents the losses. CM impedance of a CMC can be represented by the series equivalent circuit of a suppression core: the loss free inductor (L_s) is in series with the equivalent loss resistor (R_s). The following equation relates the series impedance and the complex permeability.

$$Z = R_s + j\omega L_s = j\omega L_o(\mu_s{}' - j\mu_s{}'')$$
where

$$L_o = \mu_0 \frac{A_e}{l_e} N^2$$

(1)

N: Number of Turns

A_e and l_e: Respectively the effective area and length of the choke under consideration.

The complex permeability of the material is assumed to be known from manufacturers' data sheets and/or simulation. A model of frequency dispersion of complex permeability in ferrites is proposed in [11]. The permeability spectra of ferrite materials can be described by the superposition of two kinds of resonance phenomena (domain-wall resonances and gyromagnetic spin rotation). They depend on the characteristic dispersion parameter and their values can widely differ from one material to another. Table **2** lists the values for two sintered materials. Fig. **11** represents the complex permeability of a sintered MnZn ferrite. The real and imaginary parts of the complex permeability can be modelled with the following equations:

$$\mu' = 1 + \frac{\omega_d^2 \chi_{do}(\omega_d^2 + \omega^2)}{(\omega_d^2 - \omega^2)^2 + \omega^2\beta^2} + \frac{\chi_{so}\omega_s^2[(\omega_s^2 + \omega^2) + \omega^2\alpha^2]}{[\omega_s^2 - \omega^2(1 + \alpha^2)]^2 + 4\omega^2\omega_s^2\alpha^2}$$

(2)

$$\mu'' = \frac{\chi_{do}\omega\beta w_d^2}{(\omega_d^2 - \omega^2)^2 + \omega^2\beta^2} + \frac{\chi_{so}\omega_s\omega\alpha[\omega_s^2 + \omega^2(1 + \alpha^2)]}{[\omega_s^2 - \omega^2(1 + \alpha^2)]^2 + 4\omega^2\omega_s^2\alpha^2}$$

(3)

The terms (α, χ_{sc}) and (β, χ_{do}) are related to the spin component and the domain wall component respectively.

Figs. **11** and **12** present the simulated complex permeability of a sintered MnZn ferrite and the simulated CM impedance of a core made of this material respectively.

Table 2: Permeability Dispersion Parameters Of Sintered MnZn And NiZn Ferrite For Spin And Domain Wall Resonance

	Density (g/cc)	Domain Wall Component			Spin component		
		χ_{do}	f_d (MHz)	β	χ_{so}	f_s (MHz)	
MnZn Ferrite	4.9	3282	2.5	9.3*10e+6	1438	6.3	1.28
NiZn Ferrite	5.2	485	2.8	3.5*10e+6	1130	1100	161

Differential mode impedance. The value of the DM impedance is related to leakage inductances. Fluxes created by the differential mode currents do not completely cancel each other: parts of these currents leak from the choke in between the wiring system. For a more accurate estimation, formulae are available in the literature [12].

Parasitic capacitances. Parasitic stray capacitance is composed of the winding capacitance (C_w, $C_w{}'$) and the turn to turn capacitances (C_{tt}, $C_{tt}{}'$). The winding capacitances tend to decrease the efficiency of the CMC while the turn to turn capacitances tends to improve the differential mode current attenuation. Intra winding capacitances are negligibly small in most applications. Detailed formulae to model these impedances are available in [7] and [8].

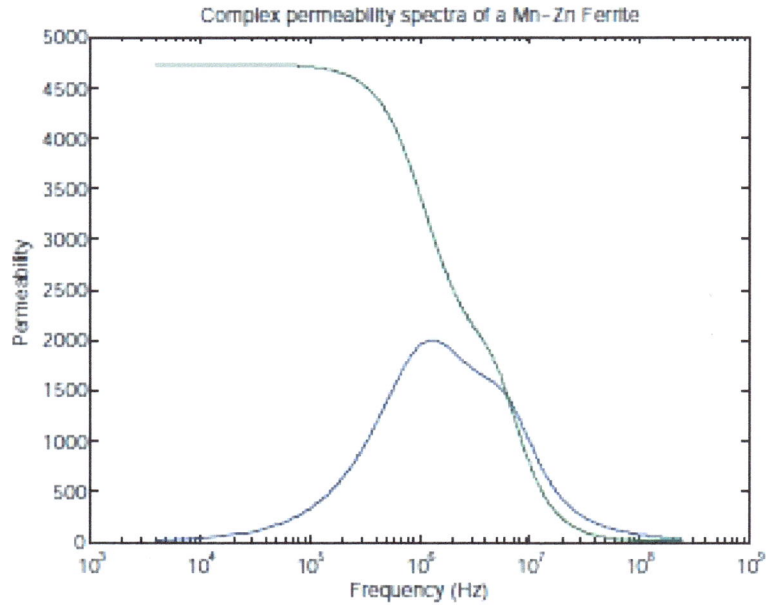

Figure 11: Complex permeability spectrum of a Sintered MnZn ferrite.

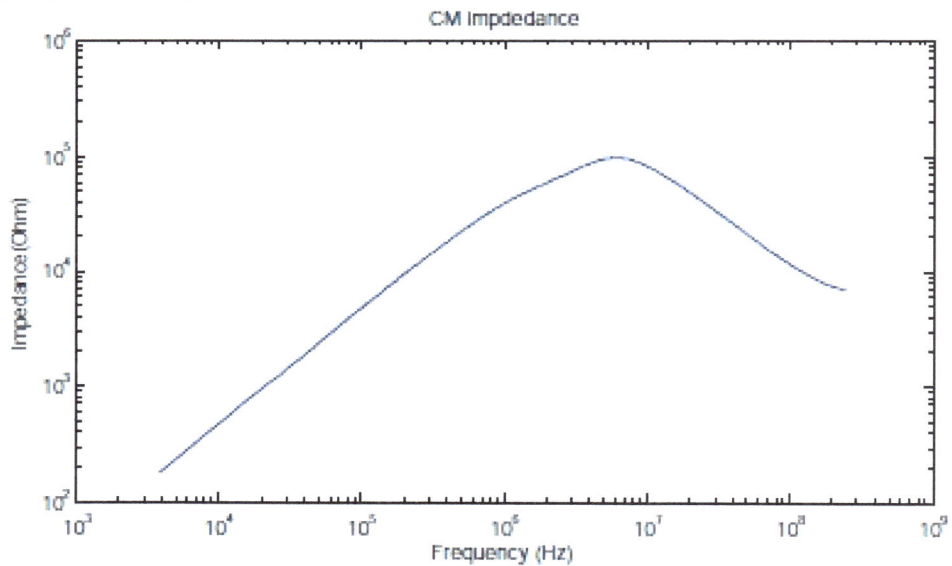

Figure 12: Simulated complex CM impedance of a MnZn core (Dimension: 25*10*5 mm, N=20).

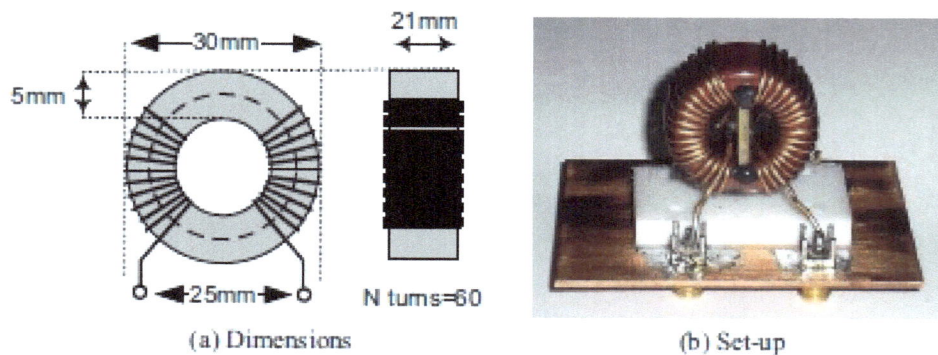

(a) Dimensions

(b) Set-up

Figure 13: CMC under test.

Measurements

Measurements setup. Measurements have been performed on a common mode choke presented in Fig. **13**. Four SMA (SubMiniature version A) connectors are connected to the four ports of the choke through a copper plate. It is then ensured that the ground of the measurement device is a reference plane which conducts the return currents. The core used as the main example in this chapter is a ferrite made of MnZn. Its characteristics have been measured with an Impedance/Gain-Phase Analyzer (HP 4194A) between 100 Hz and 40 MHz. Results are presented in Fig. **15** and **16**. This auto balancing bridge method [9] offers wide frequency coverage and is specially adapted to the ground device measurements. The test set-up used to measure the four characteristic impedances of the choke is detailed thereafter.

Measurements of coil inductance. Coil inductance can be measured directly by connecting the instrument as shown in Fig. **14-a**. All other windings should be left open. The inductance measurement includes the effects of capacitances. If an equivalent circuit analysis function is available, individual values for inductance, resistance and capacitance can be obtained. This measurement is used to extract the value of the turn to turn capacitances.

Measurements of the leakage inductance. It can be measured by connecting the output port of the first windings to the input port of the second winding to reproduce the circuit as seen by the differential mode current. The measurement is described in Fig. **14-b**.

Inter winding capacitance measurement. It is the capacitance between the two coils. It is measured by connecting one side of each winding to the instrument as shown in Fig. **14-c**. The other sides are left open.

Measurement of the common mode impedance. The inputs and the outputs windings are connected together and reproduce the circuit as seen by the common mode current, as shown in Fig. **14-d**.

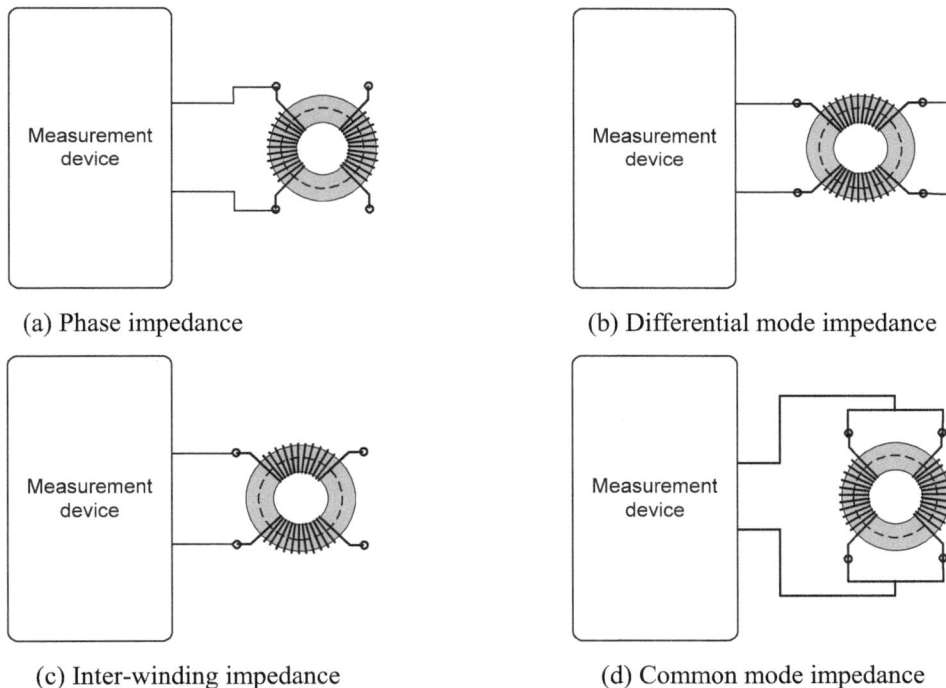

(a) Phase impedance   (b) Differential mode impedance

(c) Inter-winding impedance   (d) Common mode impedance

Figure 14: Impedances measurement.

Measurements results. The choke under consideration is bonded with a copper wire with a diameter of 0.99 mm. Its insulation in polyurethane has a thickness of 12 μm. The choke can be used with a rated current of 5A and nominal voltage of 250 Vac. The measured impedances are presented in Figs. **15** and **16**.

The equivalent circuit of the measured impedance of the common mode choke can be extracted from the measurement results shown in Figs. **15** and **16**. The equivalent circuits of the measured impedances are shown in Fig. **18**:

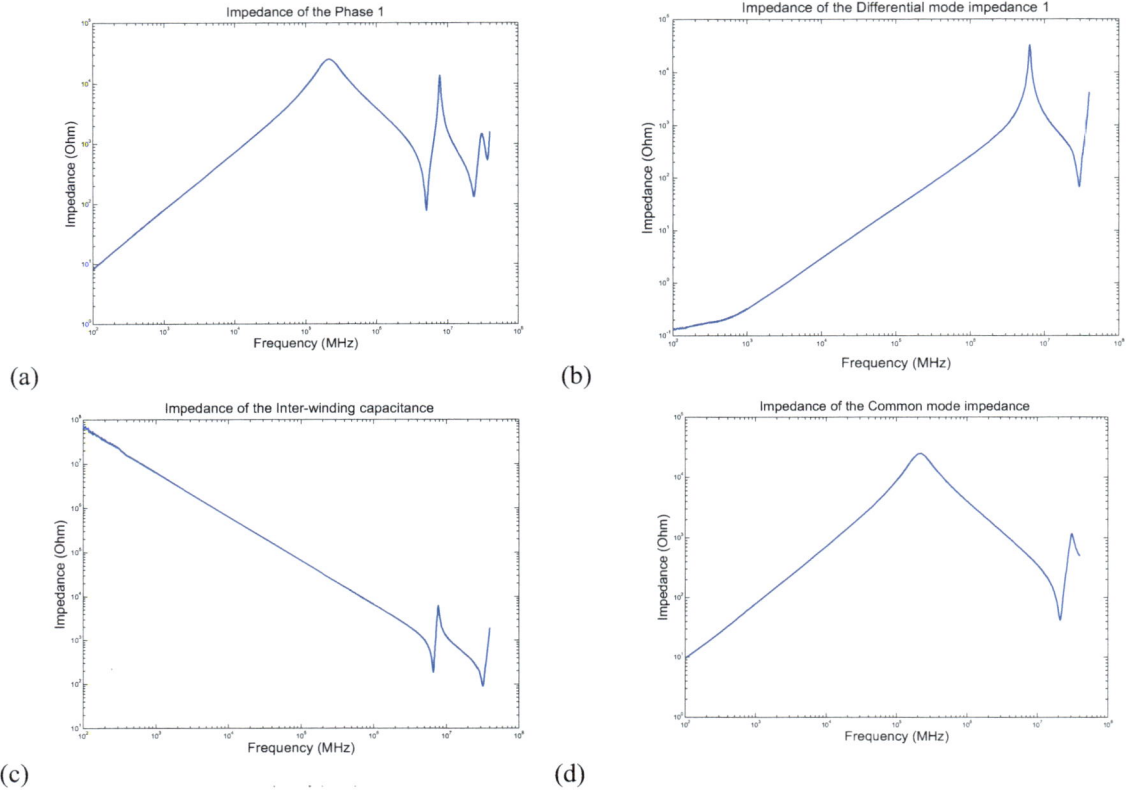

Figure 15: Measured impedances of the common mode choke under test.

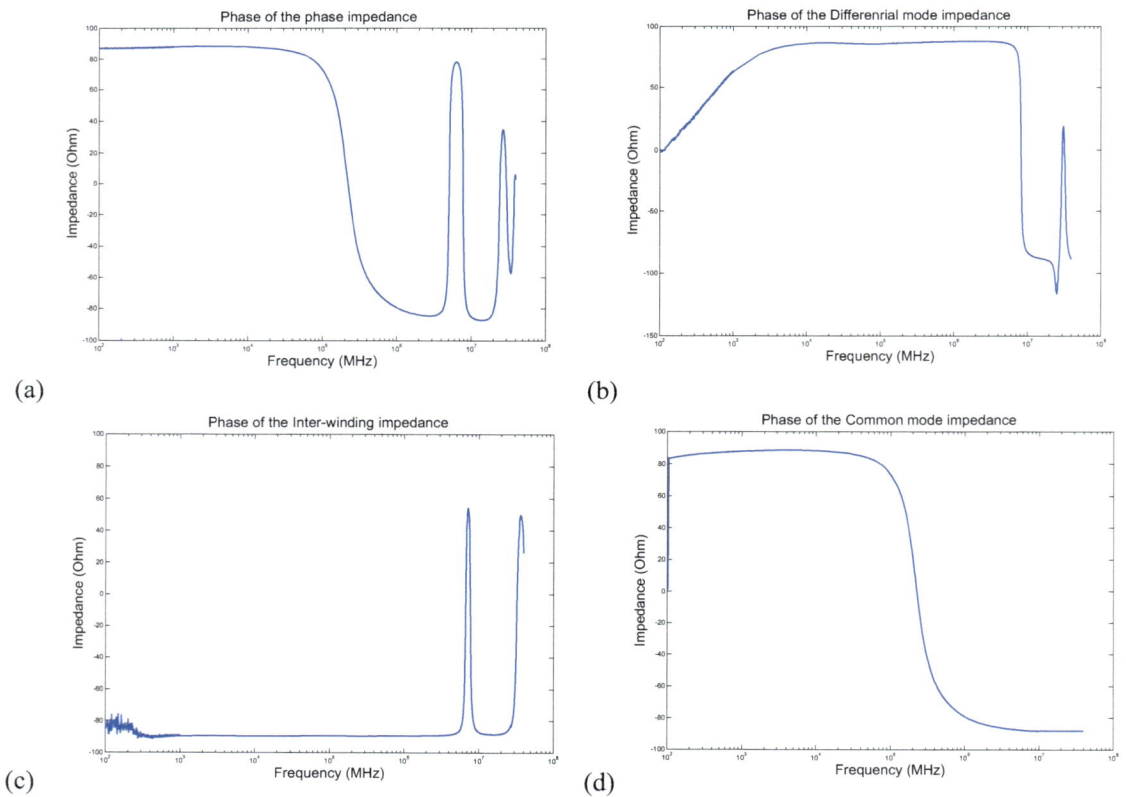

Figure 16: Measured phases of the common mode choke under test.

- The measurement of the leakage inductance (Fig. **15-b**) is independent of the inter-winding capacitances which is short circuited. The common mode fluxes generated by the windings cancel each other in this configuration. The leakage inductance resonate with the parallel turn to turn capacitance on each side of the choke This resonance can be used to extract the total value of the turn to turn capacitance which consists of two capacitances in series in this measurement configuration. The value of each turn to turn capacitance is double of the extracted value.

- The resonance between the turn to turn capacitance and the leakage inductance is also visible in the respective phase impedance in the Fig. **15-a**. It suggests that the common mode impedance is in series with half the differential mode impedance. The phase impedance remains indeed identical to the common mode impedance before this resonance. It shows that the common mode impedance is not in parallel with the turn to turn capacitance which is then only parallel with half the leakage inductance. An explanation of this structure is that the leakage flux within the wiring system interacts only with the turn to turn capacitance while the main fluxes remaining inside the choke do not.

- The anti-resonance visible in the inter-winding capacitance is related to its position in series (Fig. **15-c**) with the parallel structure formed by the differential mode impedance and the turn to turn capacitances in series with the common mode impedance.

- The measured common mode impedance is presented in Fig. **15-d** and is independent of the leakage inductance. The leakage fluxes have cancelled each other.

Fig. **17** compares the measured impedances of the choke with the modelled one. The extracted values from the measurements have been used and the good agreement between the two confirms the structure of each equivalent circuit. These results allow a clear identification of the components related to the differential mode (leakage inductance, inter-winding capacitance and turn to turn capacitances) and the common mode (common mode impedance only).

The common mode current flowing in the choke faces twice the common mode impedance in each winding as the flux on created on both sides add to each other in each winding. Both the phases are in parallel with each other as far as the common mode current is concerned. The final equivalent circuit, presented in Fig. **18**, is then equivalent to single common mode impedance. Fig. **19** shows equivalent circuits of the measured impedance.

BEHAVIORAL MODEL FOR COMMON MODE CHOKES

Criteria of a Behavioural Model of Common Mode Choke

Major criteria for behavioural modelling of common mode chokes are:

- The noise spectrum of EMI in frequency convertors is usually spread from around 10 kHz to several decades of MHz. The frequency dependence of parameters of the choke has to be taken into account to provide an accurate model even at higher frequencies.

- The model is predictive: it is assumed that designers do not have access to a prototype and experimental values. The usage of models as "scaled circuit" is not possible in this configuration. In addition to electrical information like voltages and currents, others parameters of the predictive model have to be provided by literature or manufacturers.

- The prediction method has to permit the integration of all parasitic elements. In software widely used like Pspice ®, it leads to very small time constants and to serious convergence problems.

- A common problem of common mode chokes is the magnetic saturation. Specific attention in its modelling is required.

- Time domain as well as frequency domain is needed by designers.

- The predictive model will be a 'black box' for designers. It only requires several parameters as inputs and provides outputs needed in graphs or tables.

The software Simulink ®, for its user friendly interface and its combined usage with Matlab ®, is chosen to develop the model presented.

Phase impedance modeling

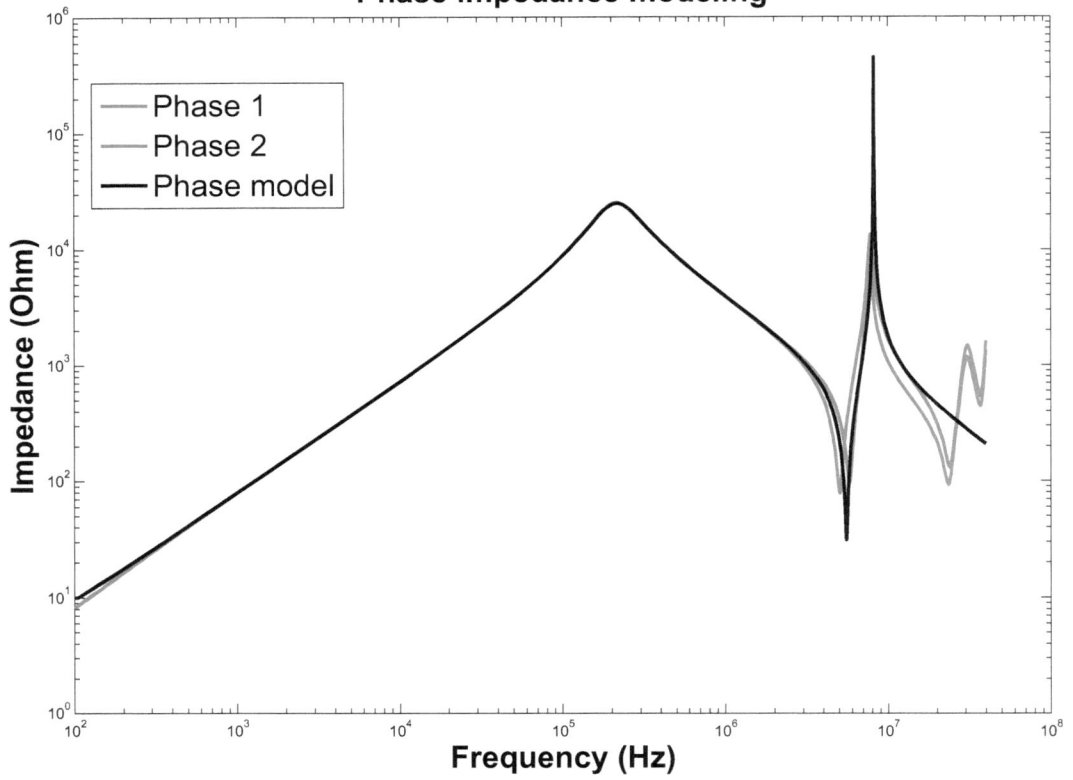

Inter-winding capacitance modelling

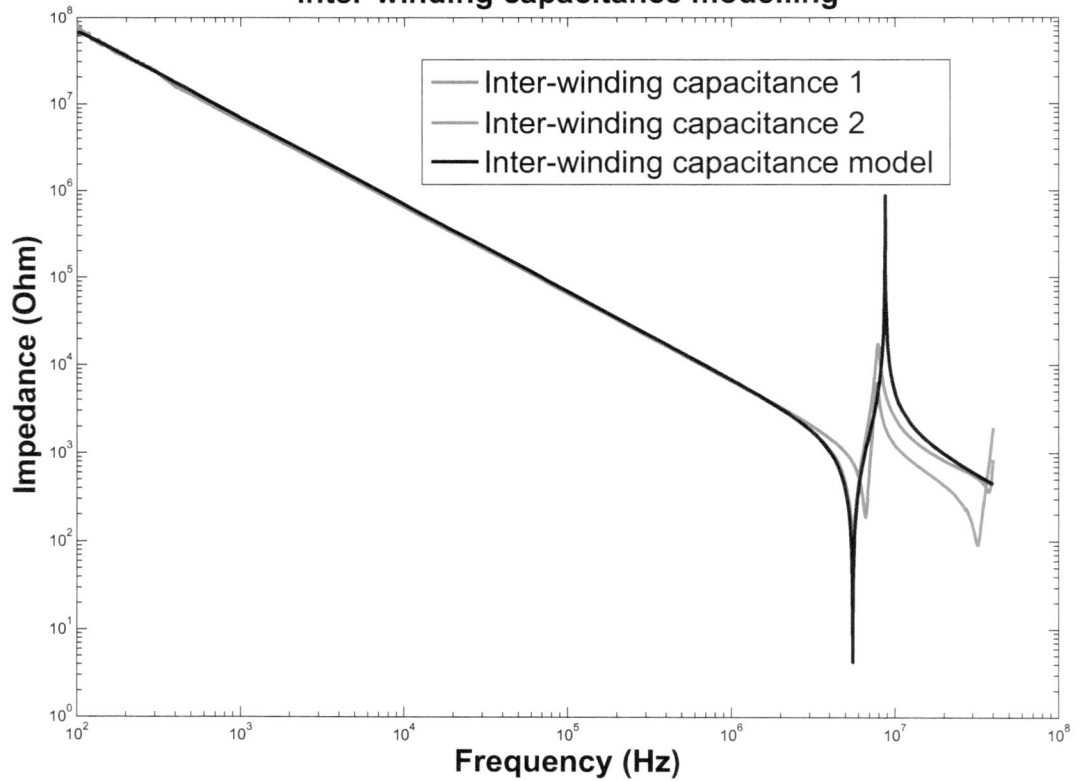

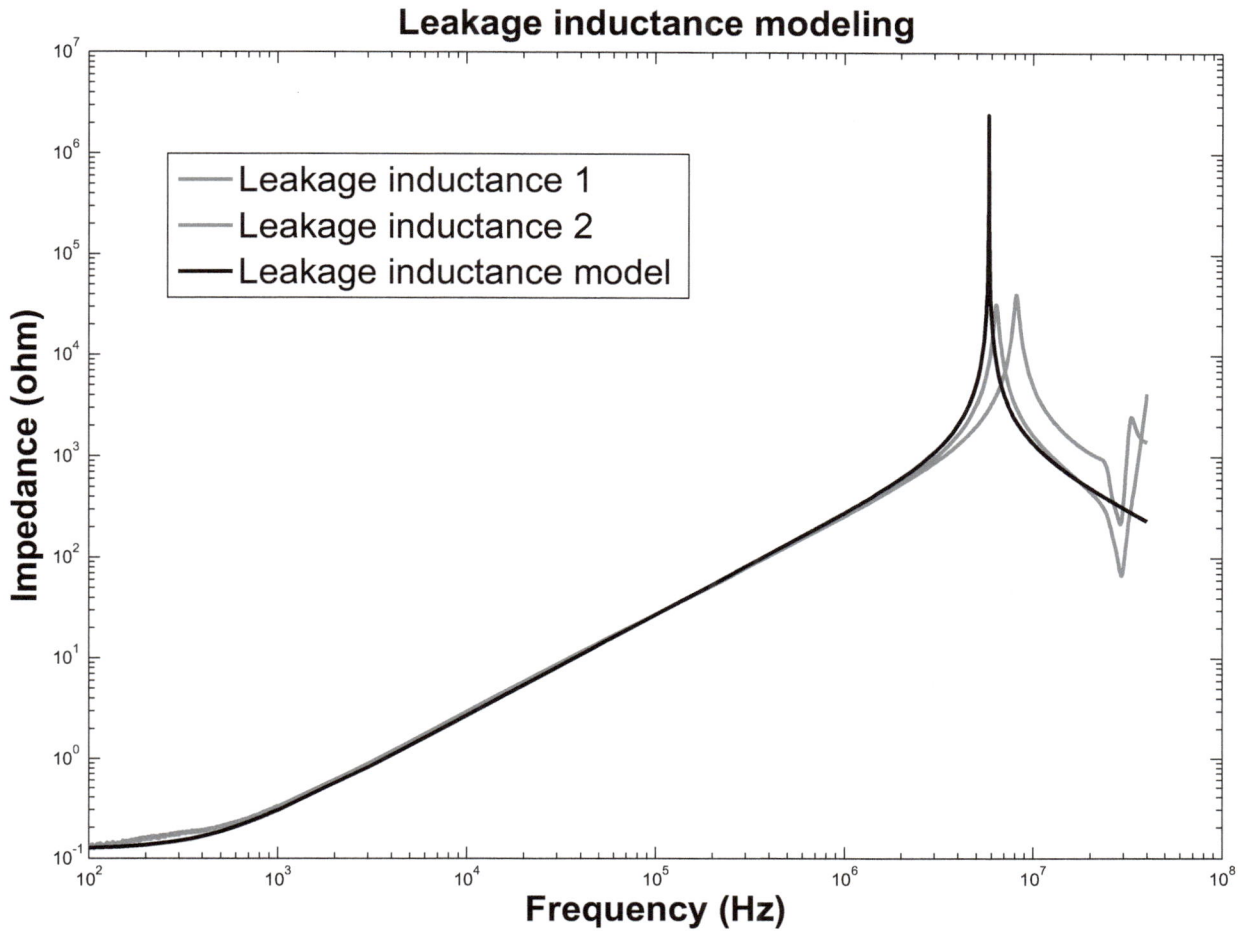

Figure 17: Comparison of the measured impedances of the choke with the modeled impedances (Extracted values).

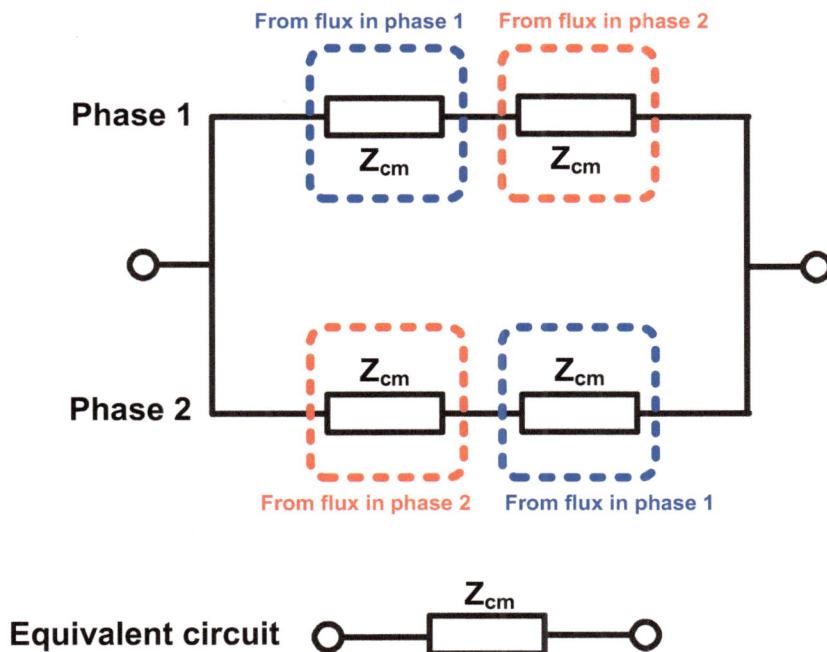

Figure 18: Equivalent common mode circuit of the choke.

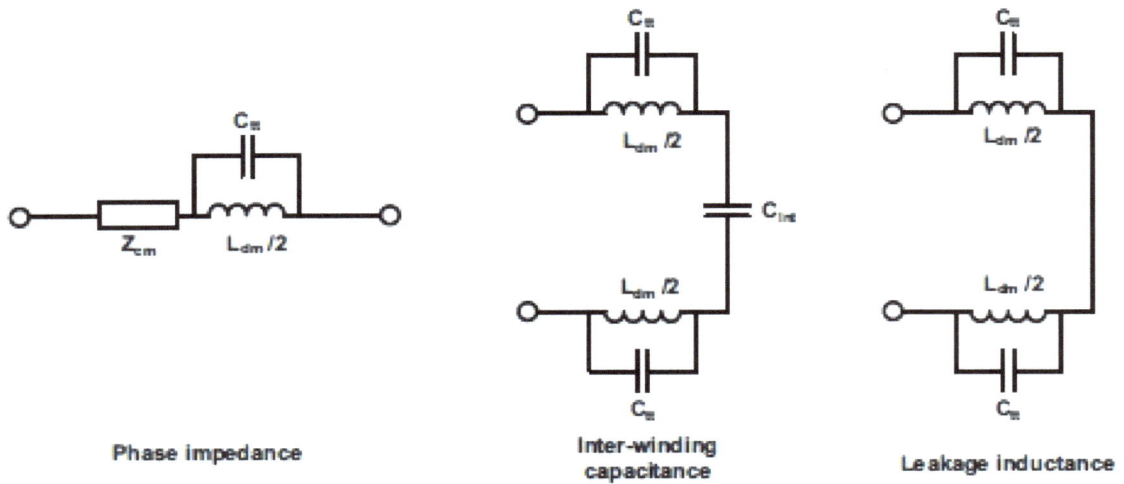

Phase impedance   Inter-winding capacitance   Leakage inductance

Figure 19: Equivalent circuits of the measured impedances.

Topology of the Model

An overview of the topology of the model in consideration is presented in Figure 20. The topology of the model is based on a translation of the common choke in its electrical equivalent circuit 'in situ' for the common mode and the differential mode. The electrical equivalent circuit includes the following impedances: common mode and differential mode impedances and parasitic capacitances (turn to turn and intra-windings). It is assumed that the common mode filter is placed as close as possible to the output of the converter.

In this condition the characterisation of the environment of the choke, performed 'in situ', should include only 2 impedances for a grounded system with 2 phases: the input asymmetrical impedance of the cable and its load, and the phase to phase impedance of this same system. The converter is assumed to be voltage regulated. The attenuation of the common mode and the differential mode currents in the cable is then only dependant on the filter and the converter environment. This 'in situ characterisation' of the performances of the common mode choke allows all the parasitic upstream of the motor drive to be included.

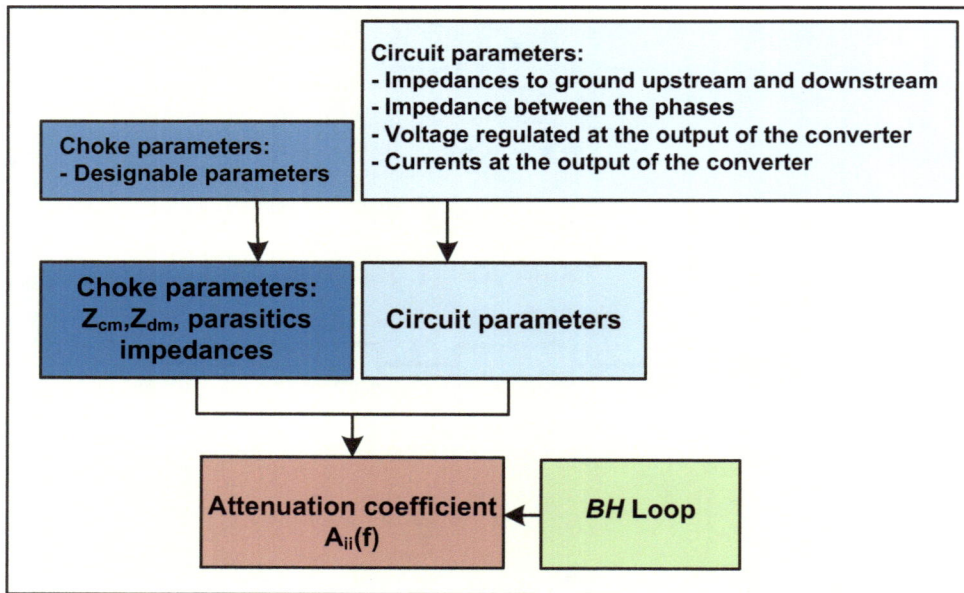

Figure 20: Topology of the behavioral model for CMC.

The designer has to find the measured or simulated currents value where the CMC will be placed later. The modification factors are first calculated in the frequency domain. If the output currents are measured in a time domain, it is important to sample the measurements properly in order to include all the high frequency characteristics of the noise.

Modification Factors Calculations

Electrical equivalent networks. Two electrical circuits are considered to characterize the attenuation of the common mode current and the differential mode current. The Fig. **21** and the Fig. **22** present the differential mode and the common mode equivalent circuit of the output of the converter with and without the choke respectively. In these circuits, Z_a refers to the asymmetrical input impedance of the cable and its motor, Z_m is the phase to phase input impedance of the motor and its cable.

These two impedances include for Z_a eventual Y-capacitors added in the common mode filter, and in the case of Z_m eventual inductors and X-capacitors added in a differential mode filter. These two impedances are also fully characterizing the circuit upstream of the converter and take into account all parasitic. The differential mode impedance Z_{dm} is formed by the leakage inductance (Z_{Ldm}) parallel to the turn to turn capacitance (Z_{CTT}) as described in the previous section and is expressed as:

$$Z_{DM} = \frac{Z_{LDM}.Z_{Ctt}}{Z_{Ctt} + Z_{LDM}} \tag{4}$$

In the same setup, the inter-winding capacitance (Z_{cint}) of the choke is associated with the impedance Z_m. The new impedance Z_3 is then expressed as:

$$Z_3 = \frac{Z_m.Z_{Cint}}{\left(Z_m + Z_{Cint}\right)} \tag{5}$$

Currents I_{DM1} and I_{CM1} are the CM and DM currents initially flowing at the output of the frequency converter. Currents I_{DM2} and I_{CM2} are the CM and DM currents at the output of the frequency converter when the common mode choke is placed in the circuit. The attenuation of common mode current and differential mode currents is of interest in this study.

Expression of the Modification Factor for the Differential Mode Current

In the setup described in Figure 21, the differential mode currents I_{DM1} and I_{DM2} can be expressed as followed:

$$\begin{cases} V_{dm} = I_{DM1}.Z_m \\ V_{dm} = I_{DM2}.(Z_3 + Z_{dm}) \end{cases} \tag{6}$$

The differential mode current modification factor Att_{dm} is:

$$\begin{cases} I_{DM2} = Att_{dm}.I_{DM1)} \\ Att_{dm} = \frac{Z_m}{(Z_3 + Z_{dm})} \end{cases} \tag{7}$$

Expression of the Modification Factor for the Common Mode Current

In the setup described in Fig. **22**, the common mode currents I_{CM1} and I_{CM2} can be expressed as followed:

$$\begin{cases} V_{cm} = I_{CM1}.Z_a \\ V_{cm} = I_{CM2}.(Z_a + Z_{cm}) \end{cases} \tag{8}$$

The common mode current modification factor Att_{cm} is:

$$\begin{cases} I_{CM2} = Att_{cm}.I_{CM1} \\ Att_{cm} = \dfrac{Z_a}{(Z_a + Z_{cm})} \end{cases}$$

(9)

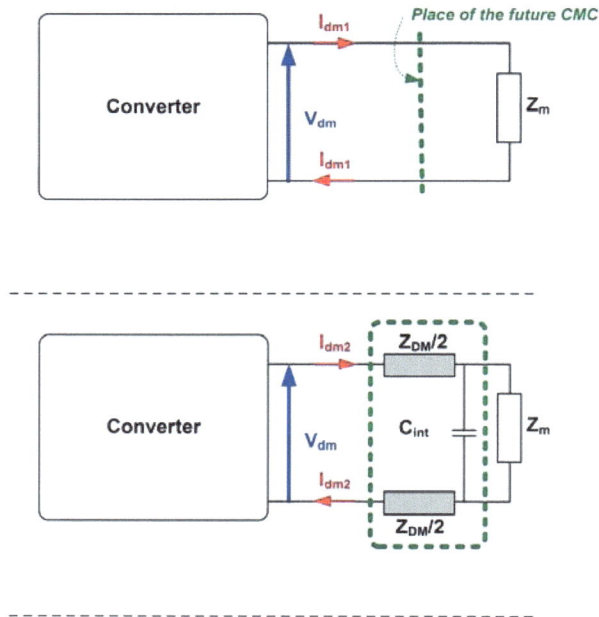

Figure 21: Differential mode equivalent circuit with and without the choke.

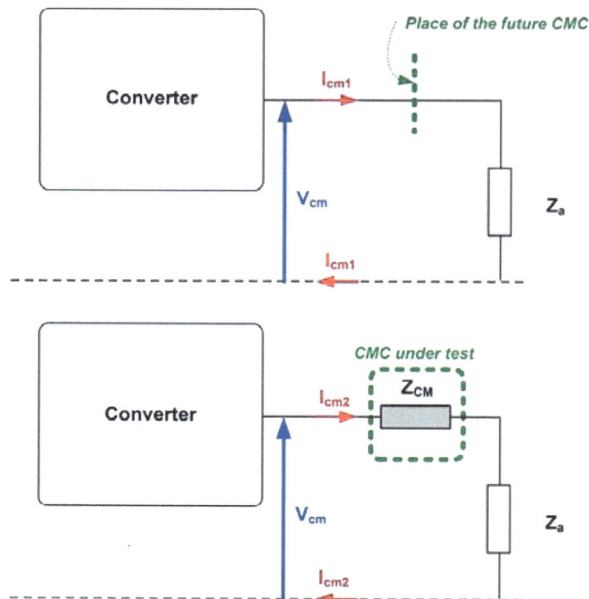

Figure 22: Common mode equivalent circuit before and after insertion of the choke.

Discussion on the Stability of the Noise Voltage

The converter has a regulated output voltage (this model can be adapted to a converter with current control). Stabilized-voltage power supplies have extremely low output impedance, often less than 1mΩ. The different techniques used in voltage control are presented in [27] and [26]. It is assumed that the common mode output noises voltage V_{cm} remains identical when the choke is connected to the circuit. In case the common mode noise voltage and/or the differential mode voltage change, these changes can be inserted in the model. Its /their variations between the two setups are then evaluated by the designer and can be introduced as:

$$Att_{dm} = \frac{Z_m}{(Z_3 + Z_{dm})}.(1 + \Delta V_{dm}) \tag{10}$$

$$Att_{cm} = \frac{Z_a}{(Z_a + Z_{cm})}.(1 + \Delta V_{cm}) \tag{11}$$

and ΔV_{cm} are the variation of common mode and differential noise voltages with of the choke respectively. These variations well are as complex numbers as the impedances and depend on the frequency. The deviation calculation introduced in the next section can also be used to study the stability of the modification factors in case these variations exist and cannot be evaluated.

Modification Factors Evaluation

Fig. **23** shows the overall model used to calculate of the modification factors. All impedances considered are complex numbers: arguments as well as phases have been included.

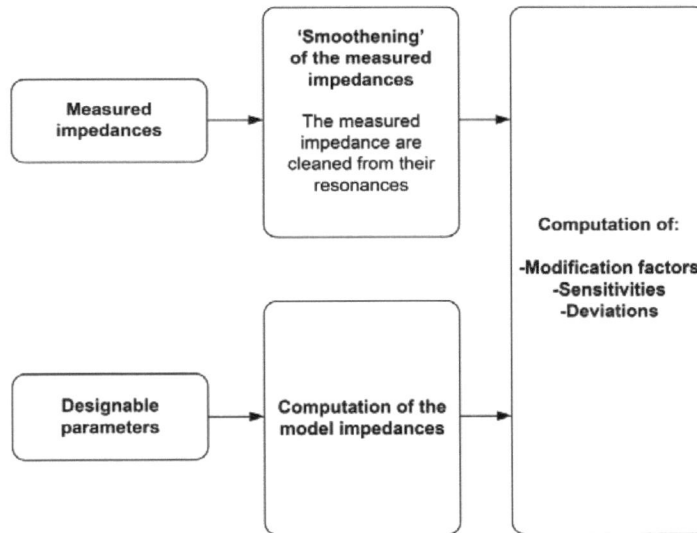

Figure 23: Overview of the model to calculate the modification factors.

Differential mode modification factor. Fig. **24** presents the choke impedances used for the differential mode current modification factor computation: the leakage inductance, the turn to turn capacitance and the input impedance between the two phases of the motor and its feeder. Four values for the capacitances have been considered between 500 pF and 2 nF. The relevance of these values is further developed in the last section of this chapter.

Two common mode chokes are considered as examples and named 'common mode chokes 1 and 2'. They are both MnZn ferrites as used in the measurements of section 0 of this chapter. The choke number 1 and 2 have 14 and 30 turns, respectively.

Impedances involved in the differential mode current attenuation
Common mode choke 2

Common mode choke 1

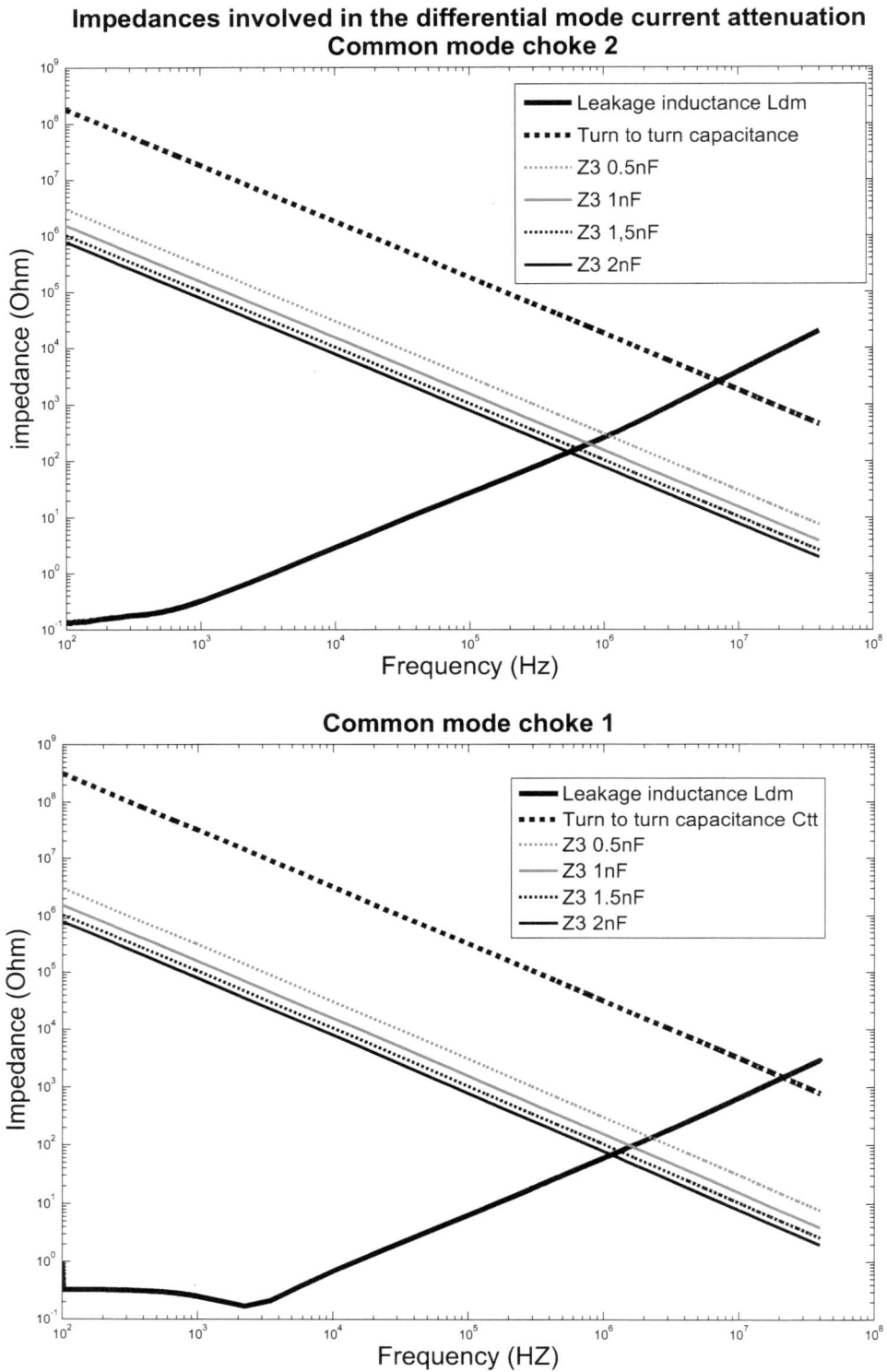

Figure 24: Differential mode impedances and variable capacitances between the phases of chokes vs. Frequency.

The Fig. **25** presents the expected differential mode modification factors with variable input cable impedances. These factors are mainly driven by the resonances of the leakages inductance with Z_m and Z_{ctt}. The resonant frequencies can be graphically evaluated at the intersection of the norm of the concerned impedances in Fig. **25**. As the frequency increases, the first resonance happens between the leakage inductance and the cable input impedance and marks the beginning of the attenuation as well as a substantial increase of the differential mode current. For the common mode choke 1 and 2 these resonances occur around 1 MHz. This increase can be a design issue for the engineers. The study of the sensitivity and the deviation of the design can be used to optimize the choke accordingly.

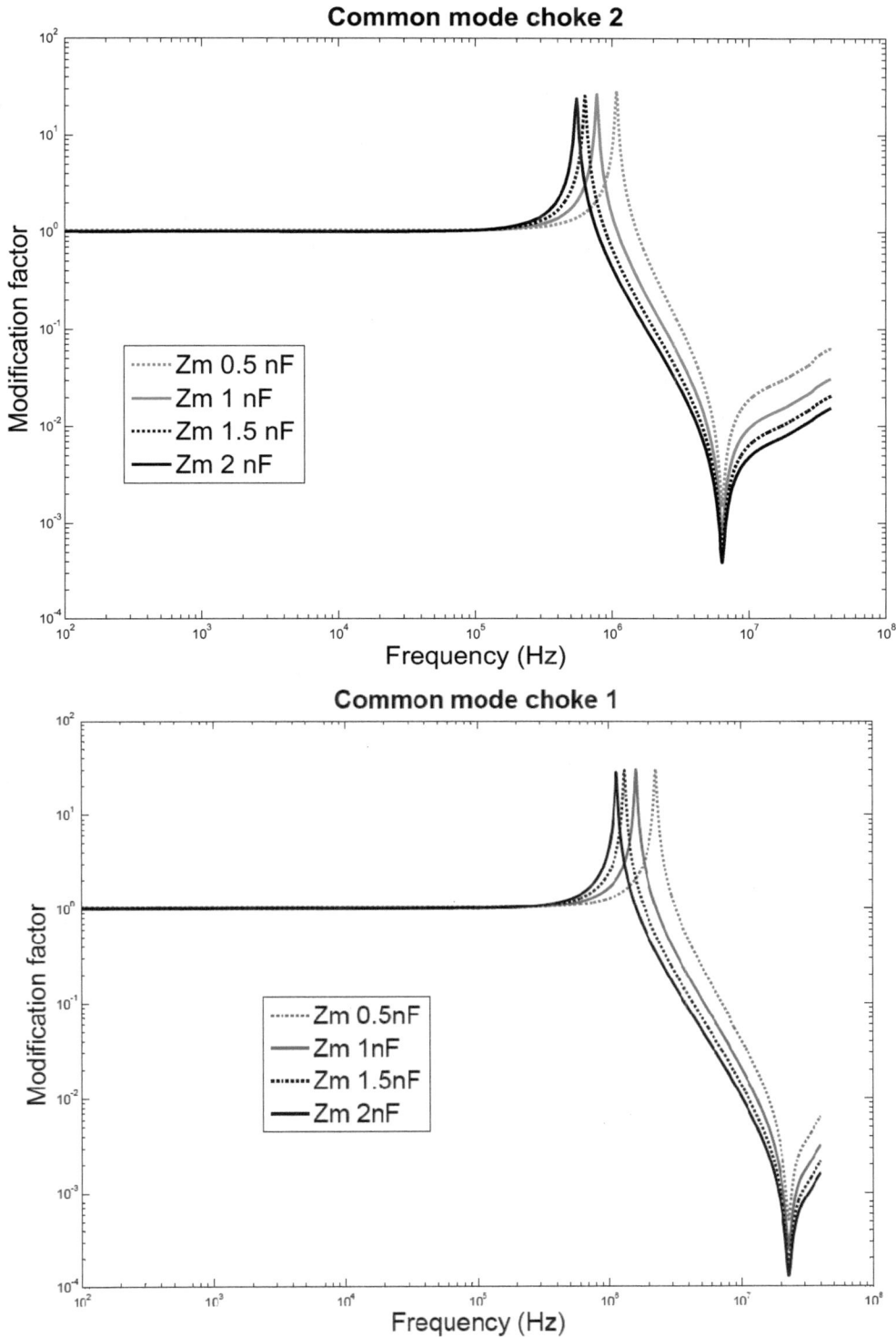

Figure 25: Differential mode modification factors with variable input cable impedance.

The maximum attenuation is reached at the resonances between the turn to turn capacitances with the leakage inductance. In the two examples the current is attenuated in the order of a thousand at the maximum attenuation. In general the attenuation of the differential mode current occurs in a narrower frequency band than the common mode one. While it is not the main purpose of the common mode choke, its prediction is however useful to forecast and prevent an eventual increase of differential mode noise in the overall filter design and/or optimize the attenuation of differential mode noise in conjunction with differential mode inductors and X-capacitors.

Common mode modification factor. Fig. 26 presents the impedances of the chokes used for the common mode current modification computation: the common mode impedance and the impedance to ground where the choke is placed. The same two common mode chokes 1 and 2 are considered.

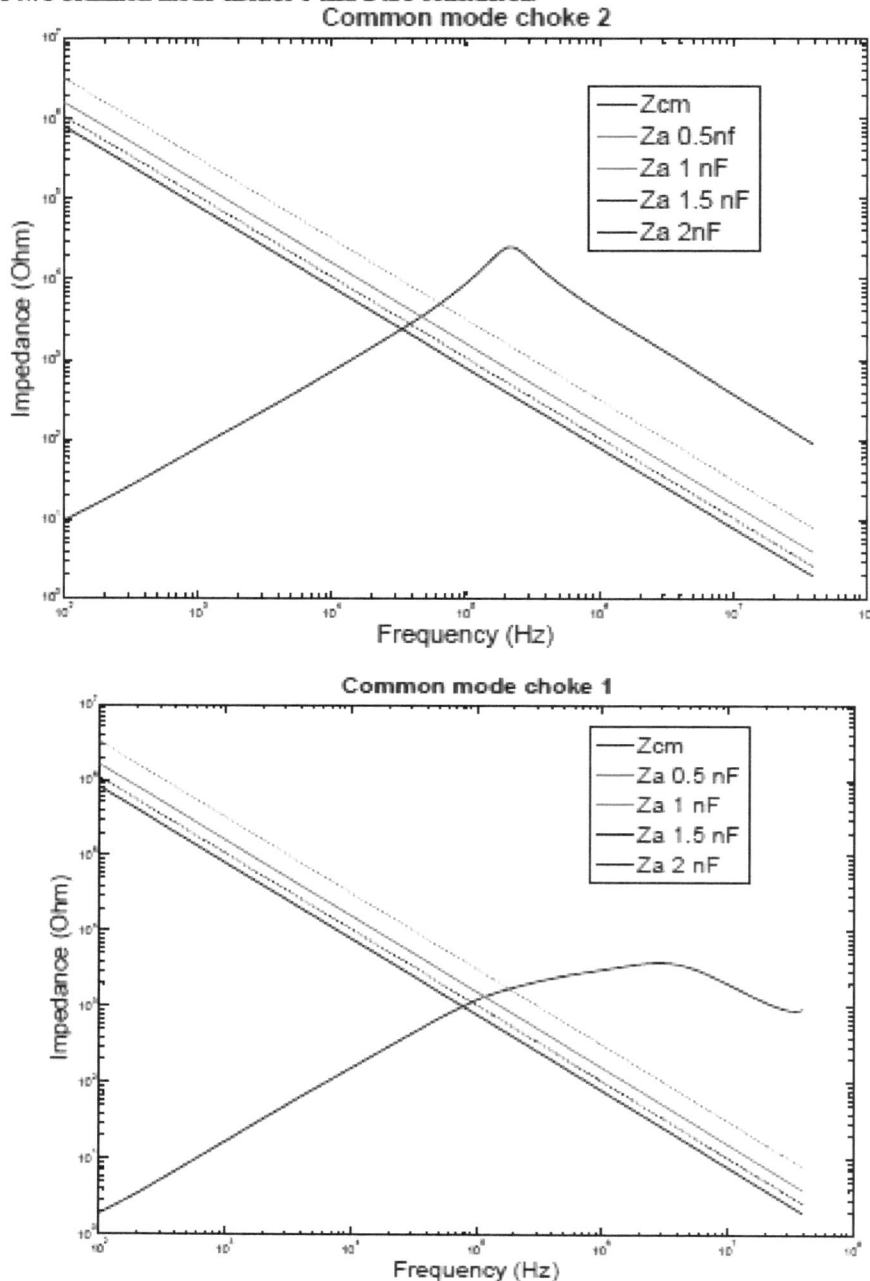

Figure 26: Common mode impedances and variable capacitances to ground of chokes vs. Frequency.

Fig. **27** presents the expected common mode modifications with variable impedance to ground. Four values of capacitance have been considered between 500 pF and 2 nF. The relevance of these choices is further developed in the last section of this chapter.

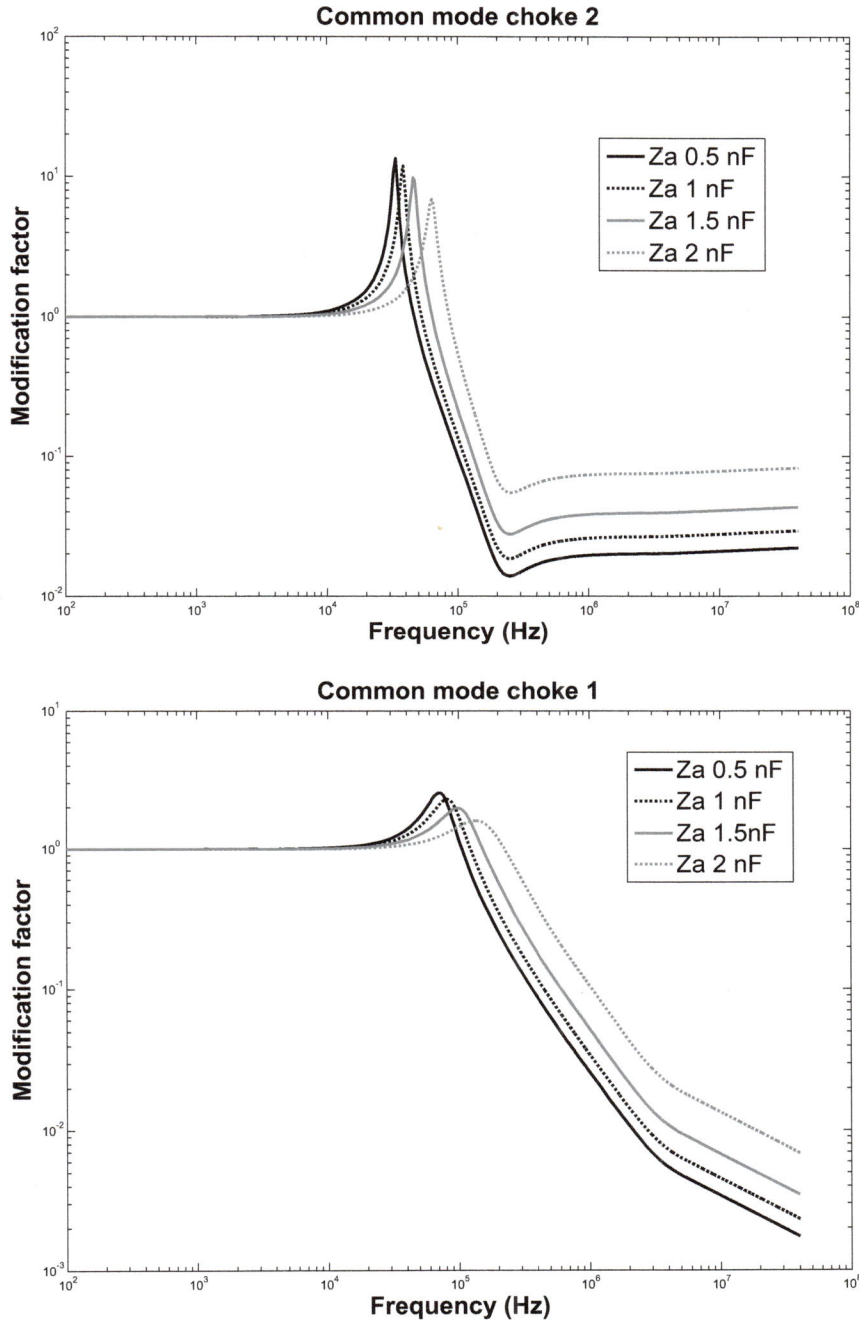

Figure 27: Common mode modification factors with variable capacitances to ground.

These modifications are mainly driven by the resonances of the impedances to ground with the common mode impedance. The resonant frequency can be graphically evaluated at the intersection of the norm of the concerned impedances shown in Fig. **26**. Likewise for increasing frequencies, the first resonance happens between the common mode impedance and the impedance to ground; and marks the beginning of the modification with a substantial increase of common mode current which can be a design issue. The study of the sensitivity and the deviation of the design can be used to design the choke accordingly. The importance of the increase of the common mode current is strongly related to the shape of the common mode impedance: a higher derivative coefficient of the impedance at the intersection point will result in a higher increase of the common mode current.

The maximum attenuation is not necessarily reached at the maximum common mode impedance value. The shape of the modification at high frequencies is related to the difference of impedance between the common mode impedance and the impedance to ground. For the common mode choke 1, this difference in impedance and the attenuation remains constant when the maximum attenuation is reached at the maximum value of the common mode impedance. For the common mode choke 2 however the difference between the common mode impedance and the impedance to ground increases after the initial resonance and the modification factor increases as well.

Saturation Modelling

To predict the performance of the common mode chokes accurately, a saturation model applicable under both steady state and transient magnetic excitations would be necessary. It is however possible in the first approximation to consider that the initial permeability remains same, as described in section 3 of this chapter. The saturation level is usually provided by manufacturer. This depends on the material, the size of the core and the number of turns.

MEASUREMENTS

Impedances Measurements

Two common mode impedances are measured:

(a)   The input common mode impedance of the load and its cable where the choke is inserted

(b)   The same impedance with the choke under test.

The common mode modification factor is obtained by dividing the input common mode impedance (a) by the one including the choke under test (b).

The Fig. **28** shows the measured input common mode impedance of the load and its cable between 20 kHz and 40 MHz frequency range. This impedance can be approximated by a capacitance of 20pF.

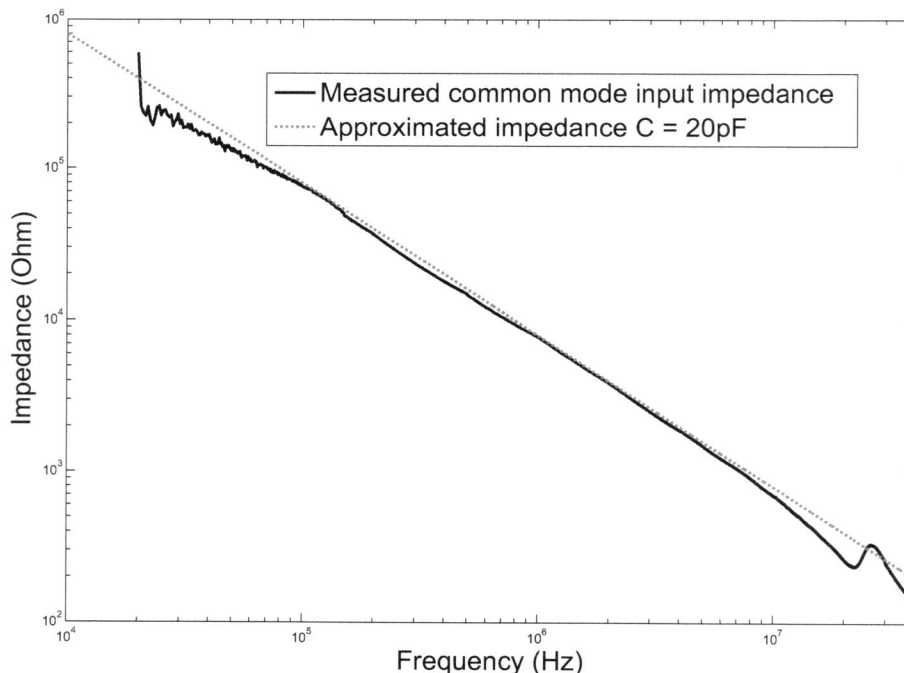

Figure 28: Input common mode impedance of the load and its cable.

The validation of the model of the common mode current modification factor is shown in Fig. **29**; the modeled attenuations are compared with the measured attenuations. The modeled attenuations are also calculated using the measured impedances of the choke 1 and 2 and approximated impedance to ground of 20pF which is shown in Fig. **28**. There is a very good agreement between the modeled and the measured values. The differences between the two curves associated with the chokes are due to the approximation of the impedance to ground with a capacitance. It is always better for the designer to use the actual impedance to ground in the simulation.

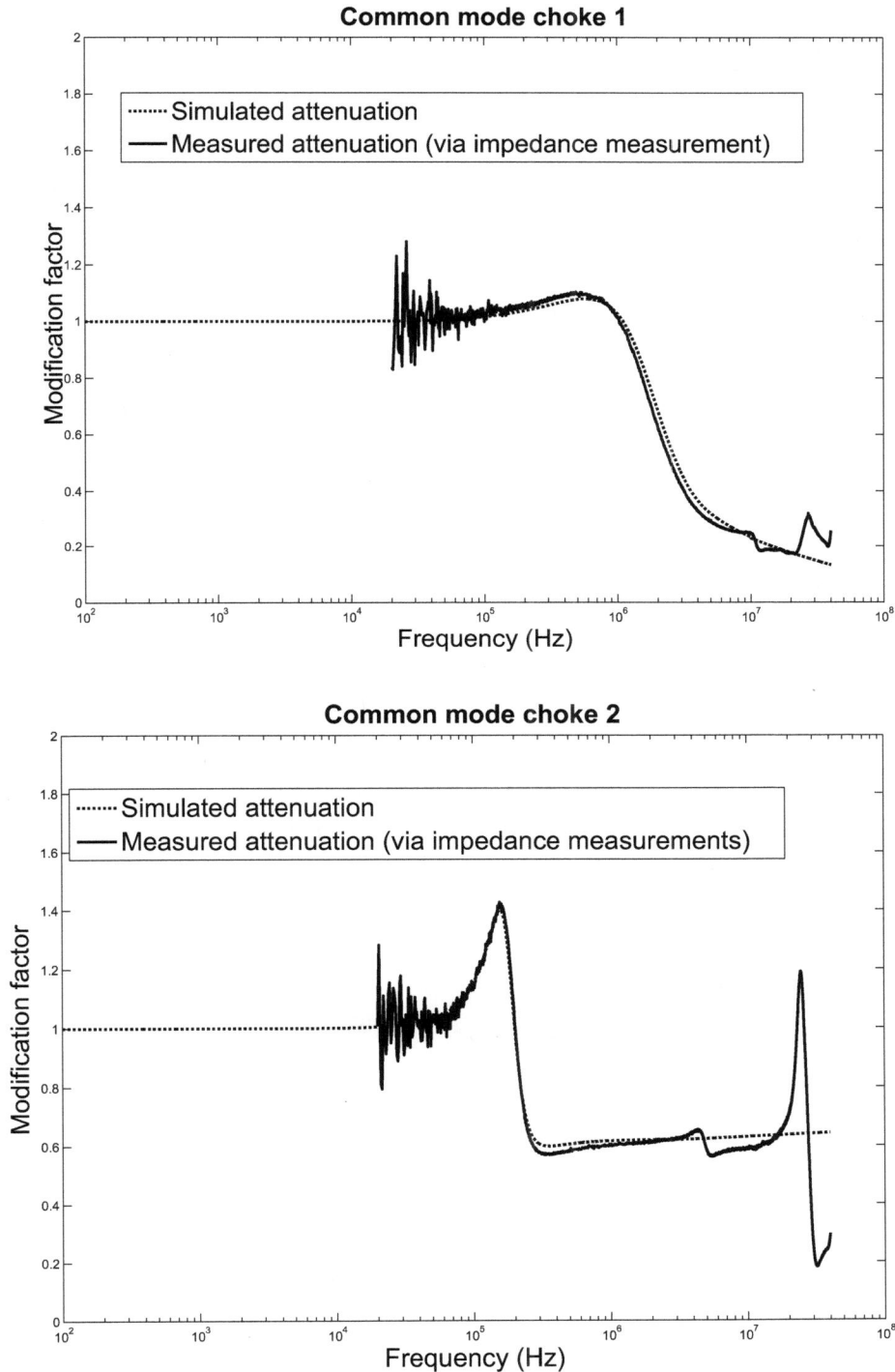

Figure 29: Modification factors: validation of the model *via* impedance measurements.

Current Modification Factor Measurements

Two common mode currents are measured:

(a) The input common mode current of the load and its cable where the choke is inserted,

(b) The same current when the choke under test is inserted.

The common mode modification factor is obtained by dividing the input common mode current (a) by the one including the choke under test (b).

The validation of the model of common mode current modification factor is shown in Fig. **30**: the modeled attenuations are compared with the measured attenuations. There is a very good agreement between the modeled and the measured values.

SENSITIVITY AND DESIGNABLE PARAMETERS

Sensitivity

Sensitivity studies provide additional insight into the behavior of the choke by understanding of how variations of parameters (for instance values of elements) influence the final performance [18], [19]. These studies are needed to take into account design variations.

The sensitivity studies can be performed at two levels. The local approach is based on a first order approximation which is mainly used to evaluate the effect of the error related to modelling or measurements of impedance and is not suitable for large variations. The normalized simplest sensitivity is the derivative function F with respect to any parameter h:

$$S_h^f = \frac{\partial \ln F}{\partial \ln h} = \frac{h}{F} \frac{\partial F}{\partial h} \tag{12}$$

When a large change occurs in the system, large change sensitivity can be considered. This study is especially relevant when an engineer wants to change the designable parameters (material, size of choke or number of turns).

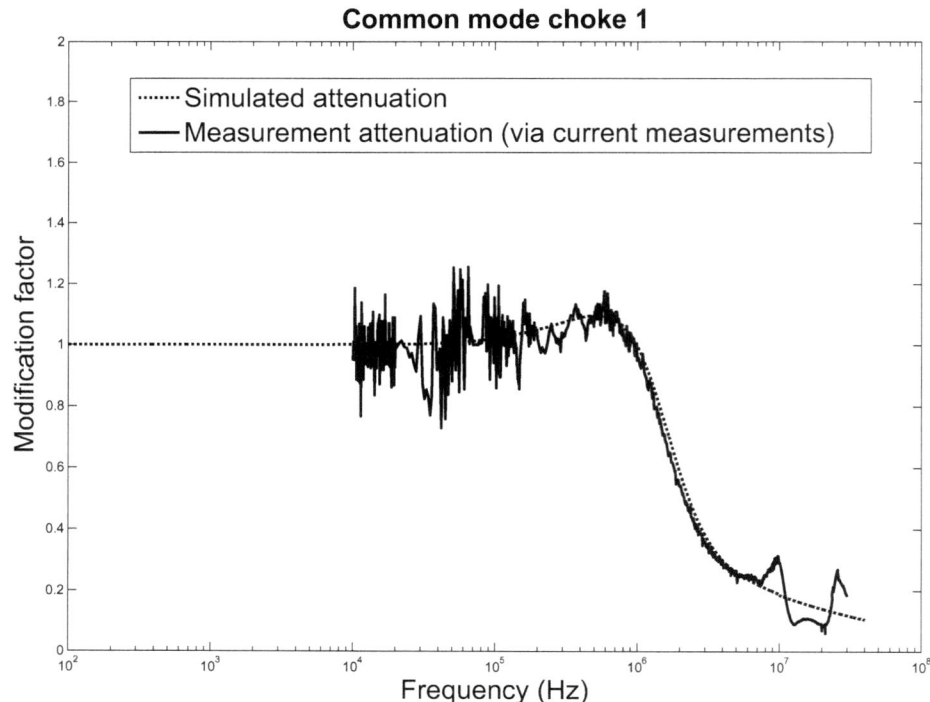

Common mode choke 1

Common mode choke 2

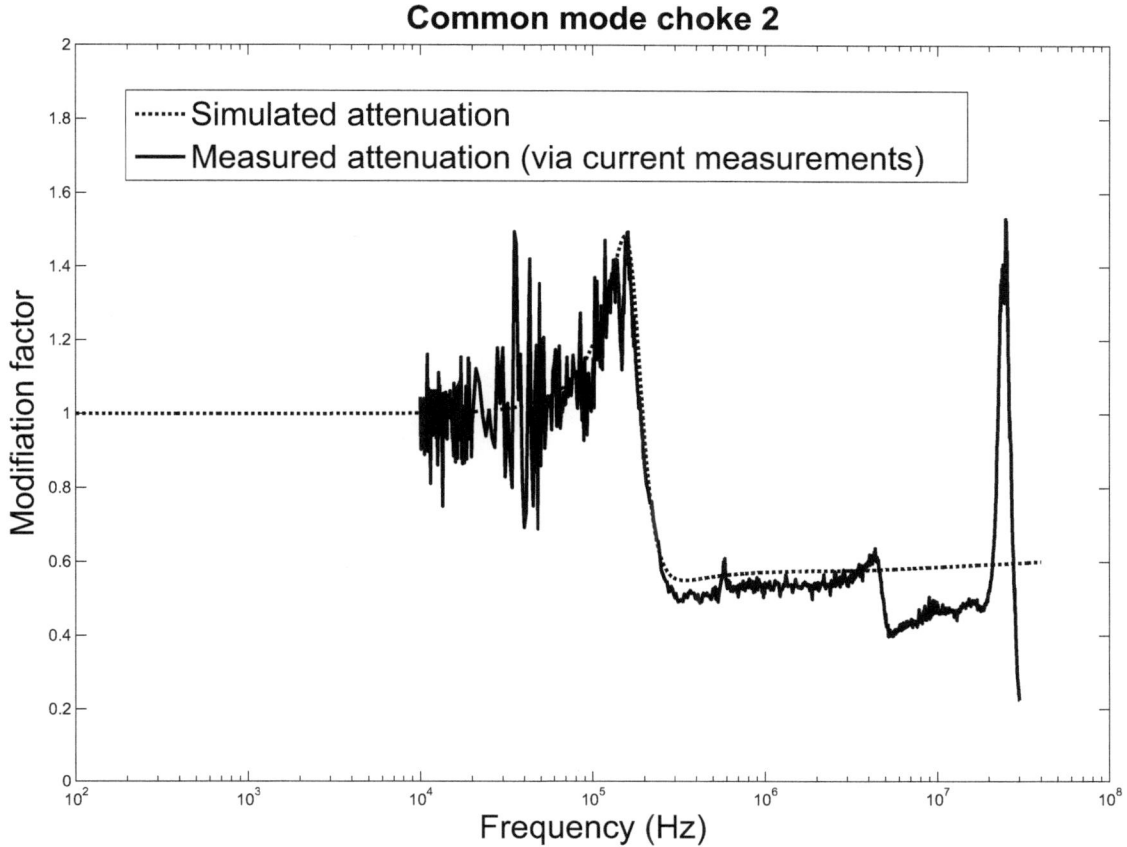

Figure 30: Modification factors: validation of the model *via* impedance measurements.

The large change sensitivity is based on a modification of the original system (the modification factors) by a low rank matrix [19].

Local sensitivity and the common mode modification factor. The common mode current modification factor is related to the common mode impedance and the impedance to the ground. The local sensitivity of the common mode impedance S_Z_{cm} and the local sensitivity of the impedance to ground S_Z_a are expressed as:

$$\begin{cases} S\_Z_{CM} = \dfrac{Z_a}{Att_{cm}} * \dfrac{\partial Att_{cm}}{\partial Z_a} = \dfrac{Z_{CM}}{(Z_a + Z_{CM})} \\ S\_Z_a = \dfrac{Z_{cm}}{Att_{cm}} * \dfrac{\partial Att_{cm}}{\partial Z_a} = -\dfrac{Z_{CM}}{(Z_a + Z_{CM})} \end{cases} \tag{13}$$

Fig. **31** presents the curves of local sensitivity of the common mode impedance and the impedance to ground for the common choke 1 and 2 presented in the previous section of this chapter. The local sensitivity has been initially expressed with the software Maple. The derived functions have then been embedded in a Simulink model. The modification factors and the choke parameters are then used as inputs in the model.

The common mode impedance presents a peak of sensitivity at the resonance between 20 kHz and 200 kHz. A slight modification of the value of the common mode impedance and/or the impedance to ground will have a predominant effect to shift the resonance frequency of the modification factor and to change its amplitude. The amplitude can be multiplied up to 10 times at the resonant frequency. After this resonance frequency, any changes in the concerned impedances compared to the modeled ones will lead to a proportional change in the modification factor. Before the resonance frequency, the sensitivity values are too low to change the attenuation significantly.

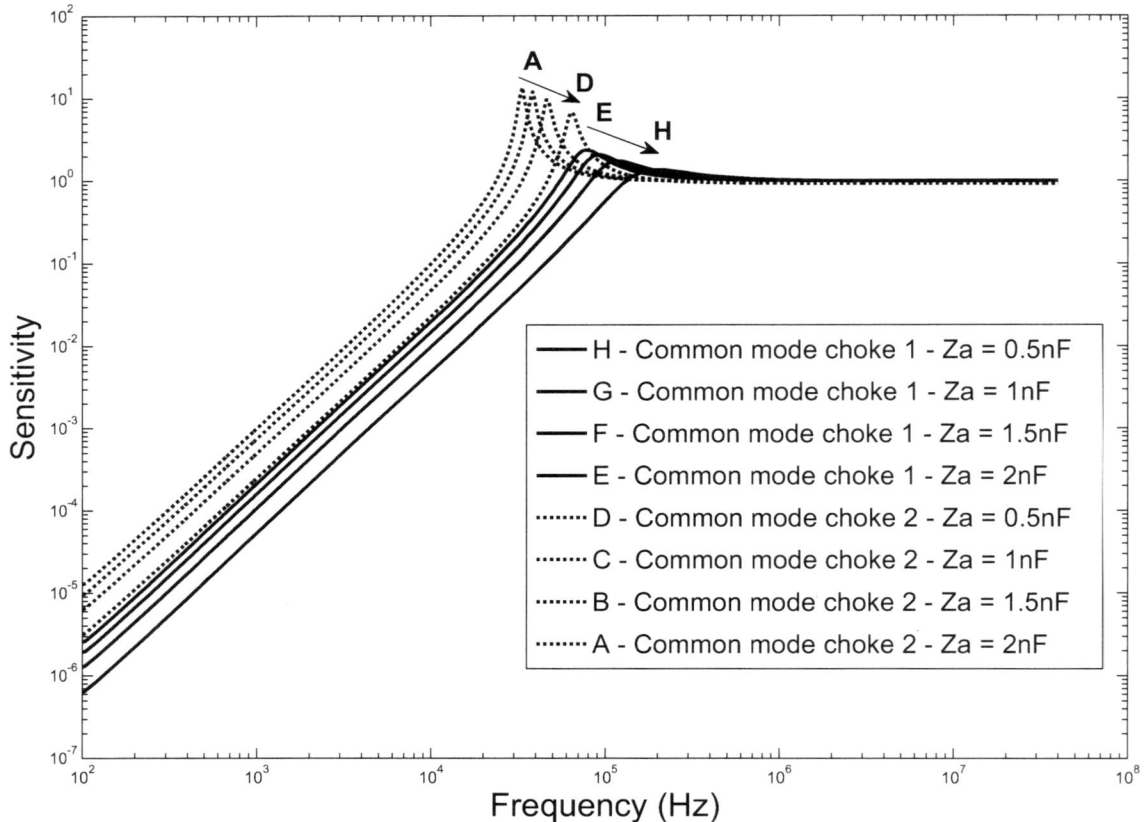

Figure 31: Local sensitivity of the impedance related to the common mode current modification factor.

Deviation

Figure 32 illustrates the upper and lower limits of the modification factor for a tolerance of 10% to 30%. A percentage of error made in the modelling of the impedances is used to evaluate two first upper and lower limits of the modification factors. This percentage has been chosen to be 10% and 30% but can be modified by the user individually for each impedance to precise complex permeability values of a material; or the tolerance of the Y capacitors can then be used to refine the range of incertitude in the modification factors. The second source of deviation is related to the electromagnetic field generated by the cable and hence not included in the measurement.

Designable Parameters: Effect of the Material

The change of material has a main effect to shift the maximum common mode impedance value. The choke is more used for its core loss properties (related to the imaginary part of the complex permeability μ'') and less for its inductive properties (related to the real part of the complex permeability μ'). The main purpose of a common mode choke is not for energy storage but it is for transformation of this energy to heat. Typically the choice of material will depend on the frequency of the noise to be filtered; a material will be efficient if the noise is located in the upper values (above 40%) of μ''.

The change of material is also an alternative when the level of saturation is reached. A nanocrystalline material is for instance a good alternative choice for a saturating choke made of iron.

Designable Parameters: Effect of the Size of the Choke

Fig. **33** presents the effect of the size of the core by comparing the impedances of two common mode chokes with two different sizes: the choke 1 (25×15×5 mm, copper wire thickness) and the choke 1-XL (35×15×7 mm, copper wire thickness 1.5mm). The choke 1 is same as already introduce in the chapter. The choke 1-XL has the same structure but a bigger core.

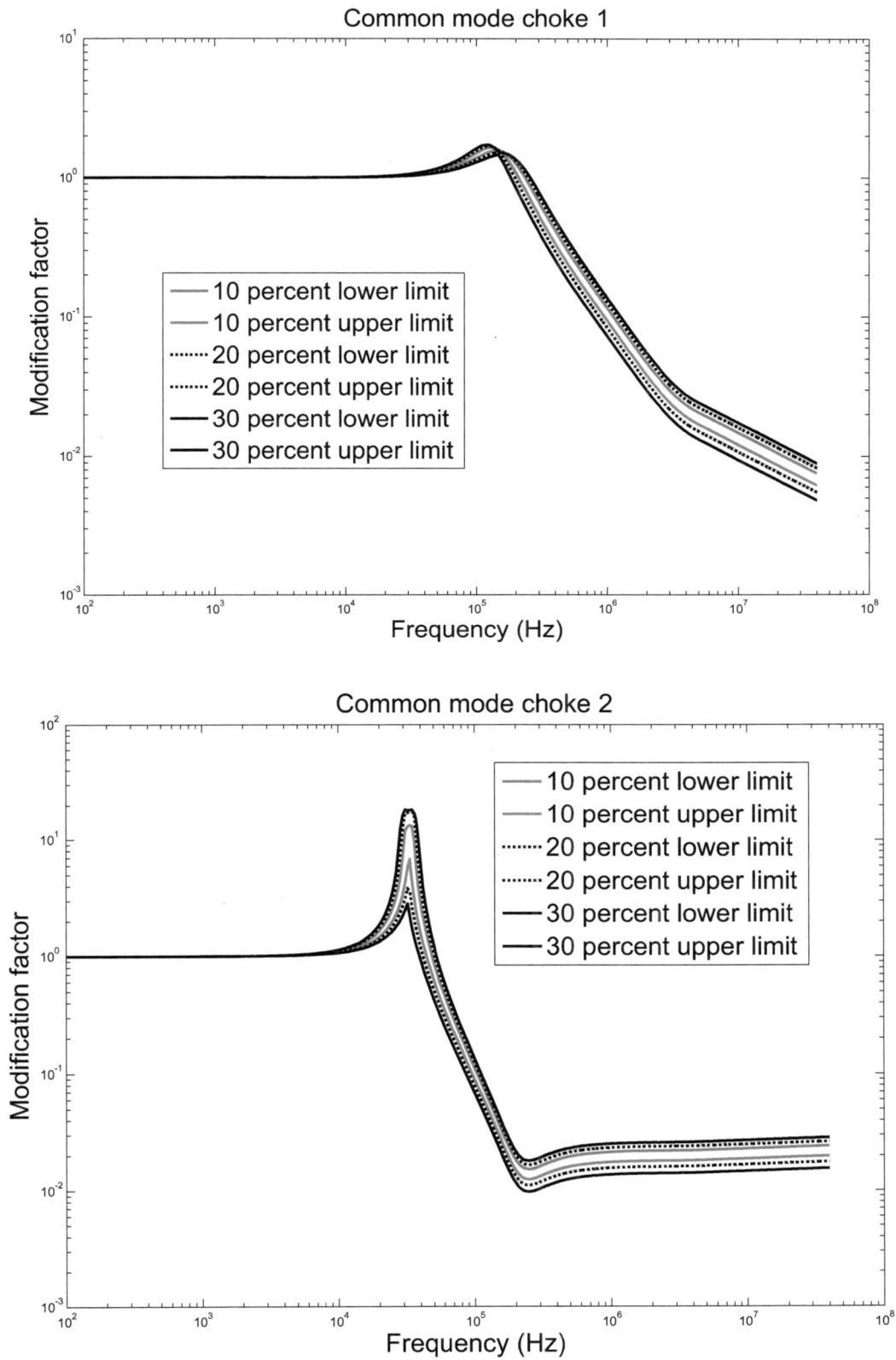

Figure 32: Upper and lower limits of the modification factor (Incertitude from 10 to 30 percent).

The turn to turn capacitance and the leakage inductance of the choke 1-XL are a bit higher due to a thicker and a longer windings but remains within the same range of value.

The increase of the size of the choke tends to shift the maximum value of the common mode impedance towards the lower frequency. This shift may or may not change the frequency at which the attention of the common mode current will start; this frequency is indeed related to the resonance between the impedance to the ground of the system with the common mode impedance. It translates graphically by an intersection of the two respective curves. If this intersection occurs where the common mode impedance is dissimilar between the two chokes, the attenuation of common mode current will then start at a lower frequency for the bigger choke. Else, this frequency remains unchanged.

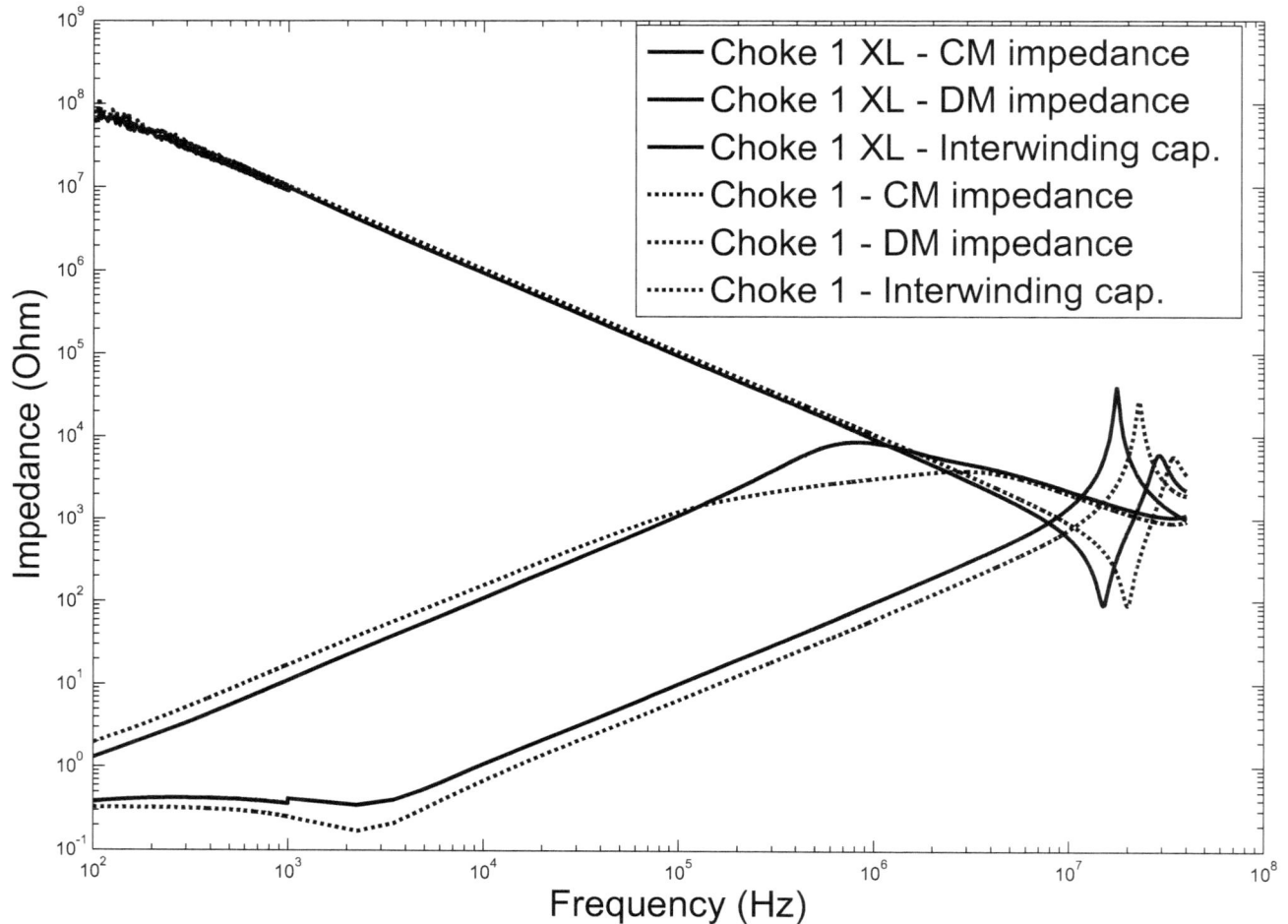

Figure 33: Impedances of two common mode chokes with different core size and same number of turns.

Designable Parameters: Effect of Number of Turns

Fig. **34** shows the influence of the number of turns on the value of the common mode impedances, the phase inductances, the turn to turn capacitances and the differential mode impedances. The core is same for all the four chokes tested.

The value of the common mode impedance increases with the square value of the number of turns. As the real part of the complex permeability is usually higher than its imaginary part, an increase the number of turns will also shift the maximum common mode impedance value towards the lower frequencies. The density of flux in the choke is increased with the number of turn and the saturation level is easier to reach.

Phase inductances

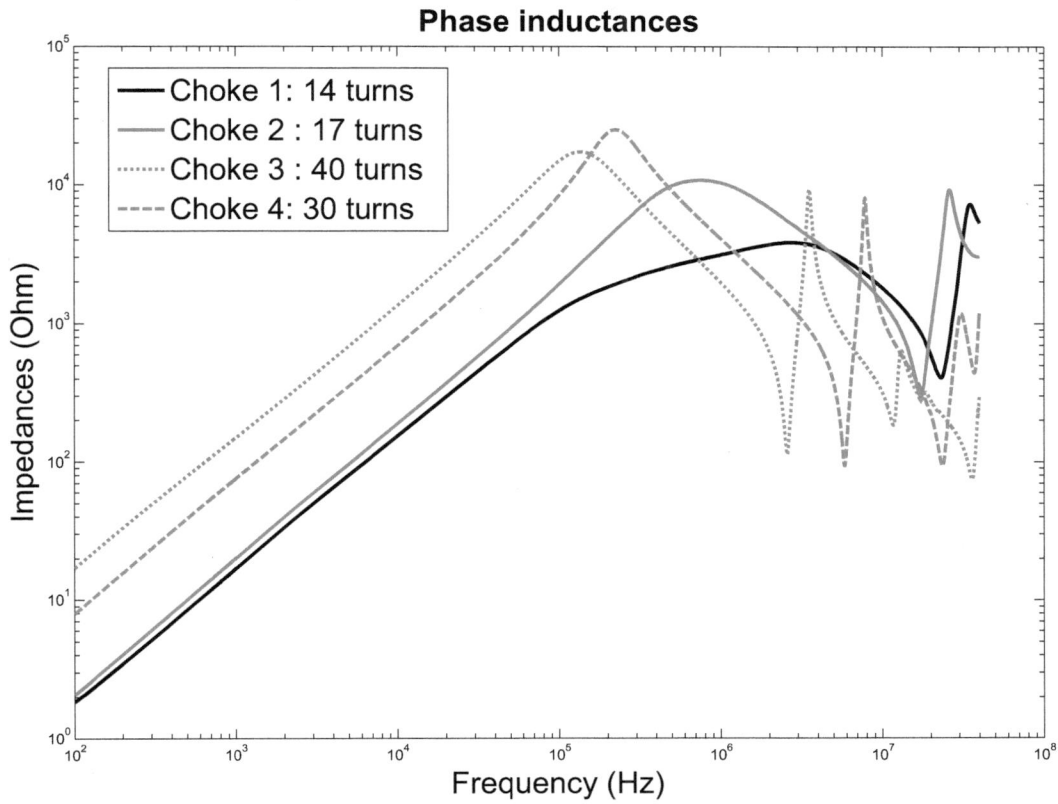

Common mode impedances

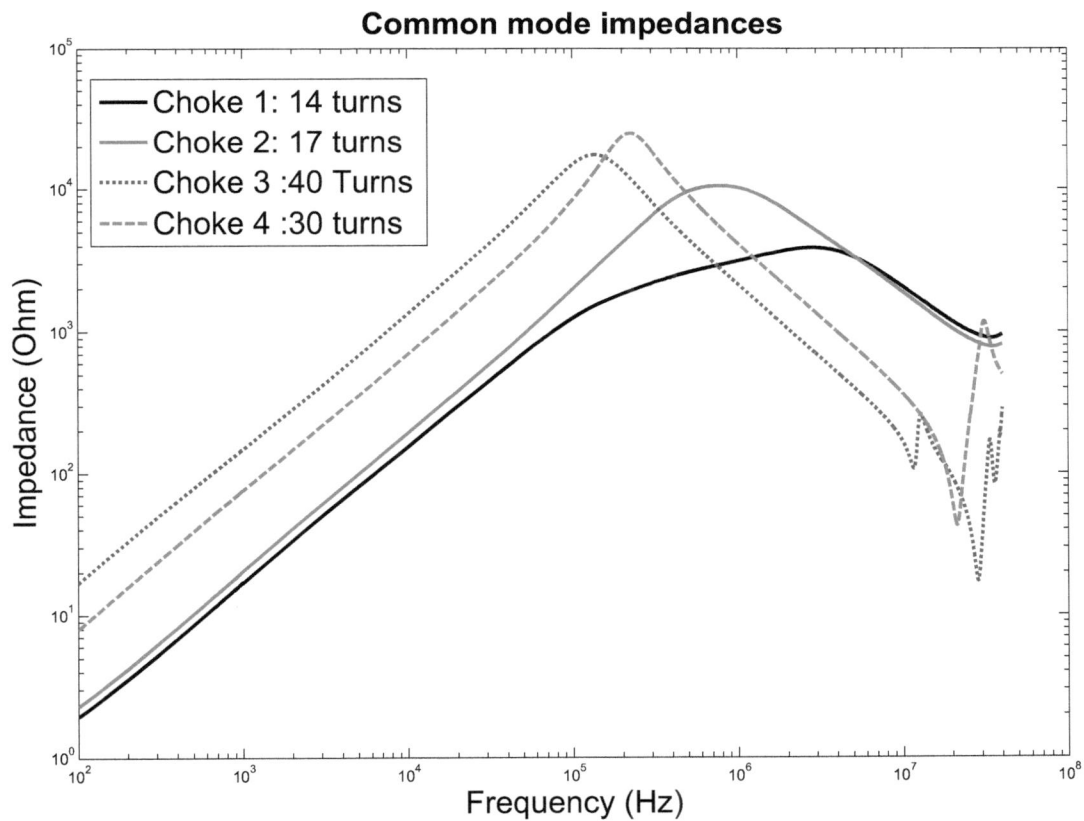

Differential mode impedances

Interwinding capacitances

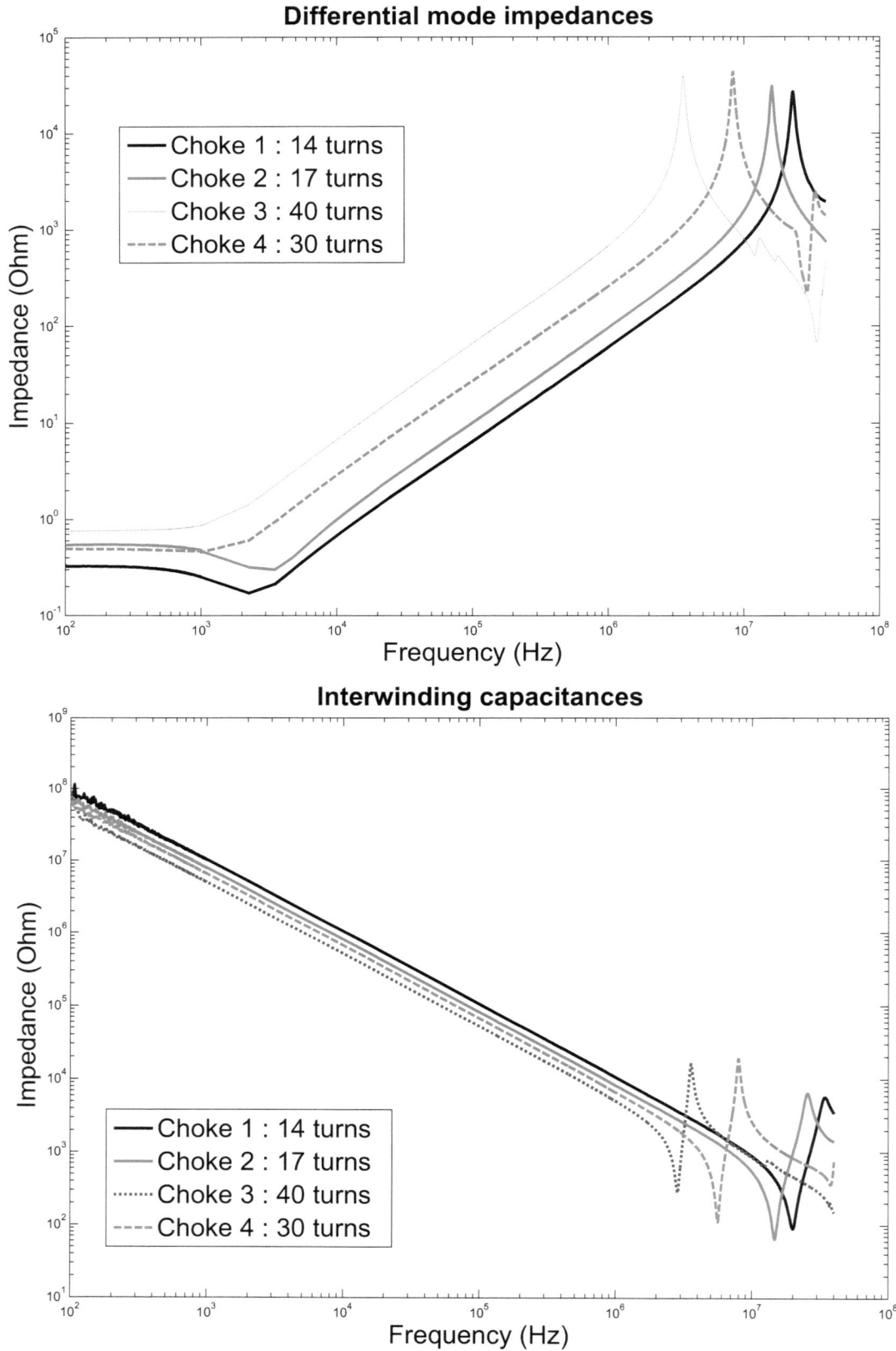

Figure 34: Influence of the number of the number of turns on: common mode impedances, turn to turn capacitances, differential mode impedances.

Designable Parameters: Effect of the Wiring System

Two kinds of windings are possible: the sectional winding and the bifilar winding. Sectional winding consists of the same number of windings placed at diametrically opposite ends on the toroidal core and bifilar winding consists of a parallel wiring around the core. The two families of windings are shown in Fig. **35**. Their respective impedances are presented in Fig. **36**. The main advantage of the sectional winding stands in the high leakage inductance which allows the filtering of the differential mode current in addition of the common mode one. The small leakage inductance provides less attenuation of high frequency differential mode current.

Designable Parameters: Effect of the Wire Dimension

The wire dimension is chosen according to the density of current in the circuit. It has a small effect at very low frequencies where the resistance of the wiring is the main contribution to the impedances. This resistance is visible in the measurement of the leakage inductance at low frequency.

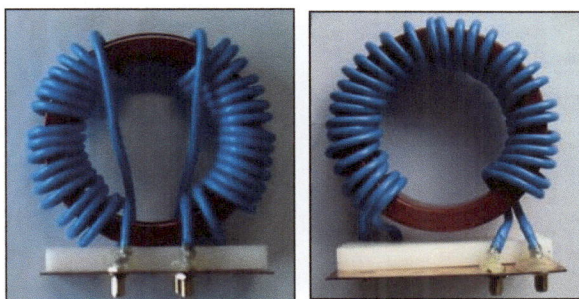

Figure 35: (a) Sectional winding and (b) bifilar winding of a common mode choke.

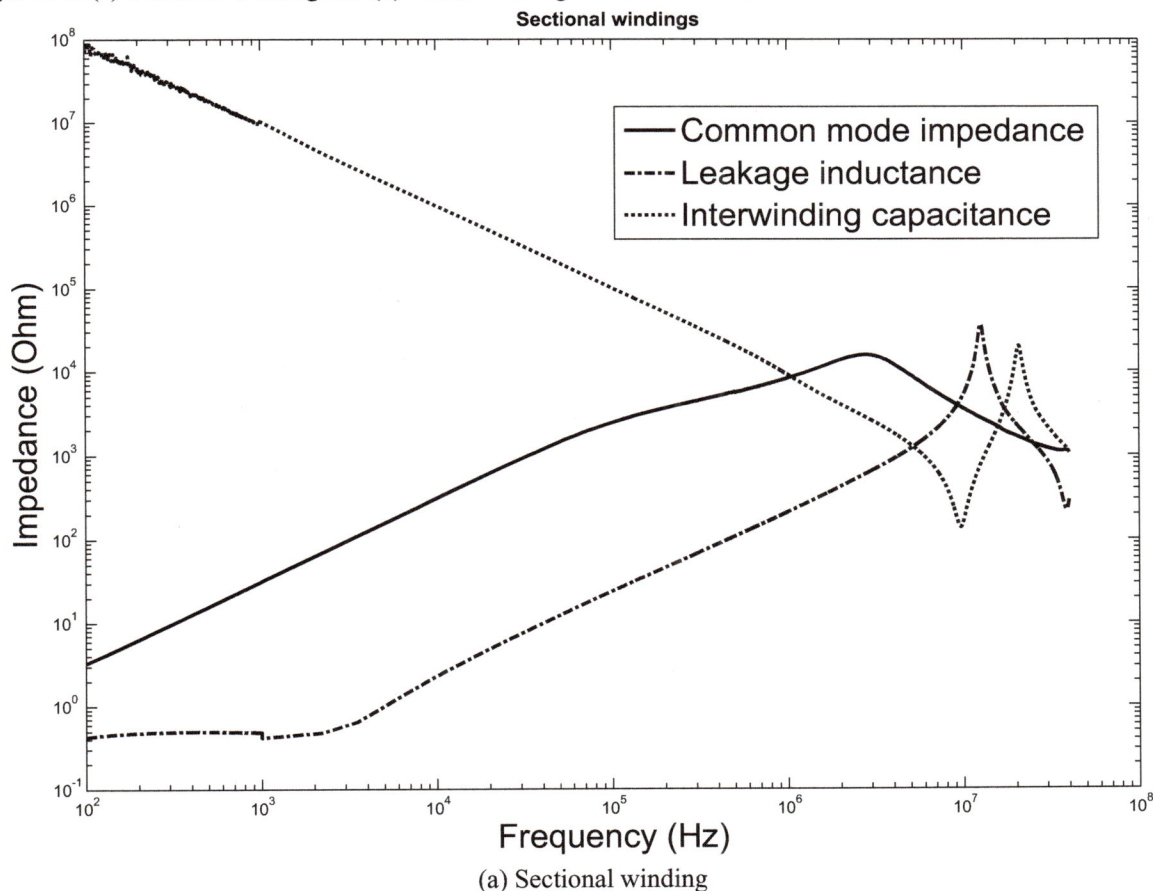

(a) Sectional winding

Bifilar windings

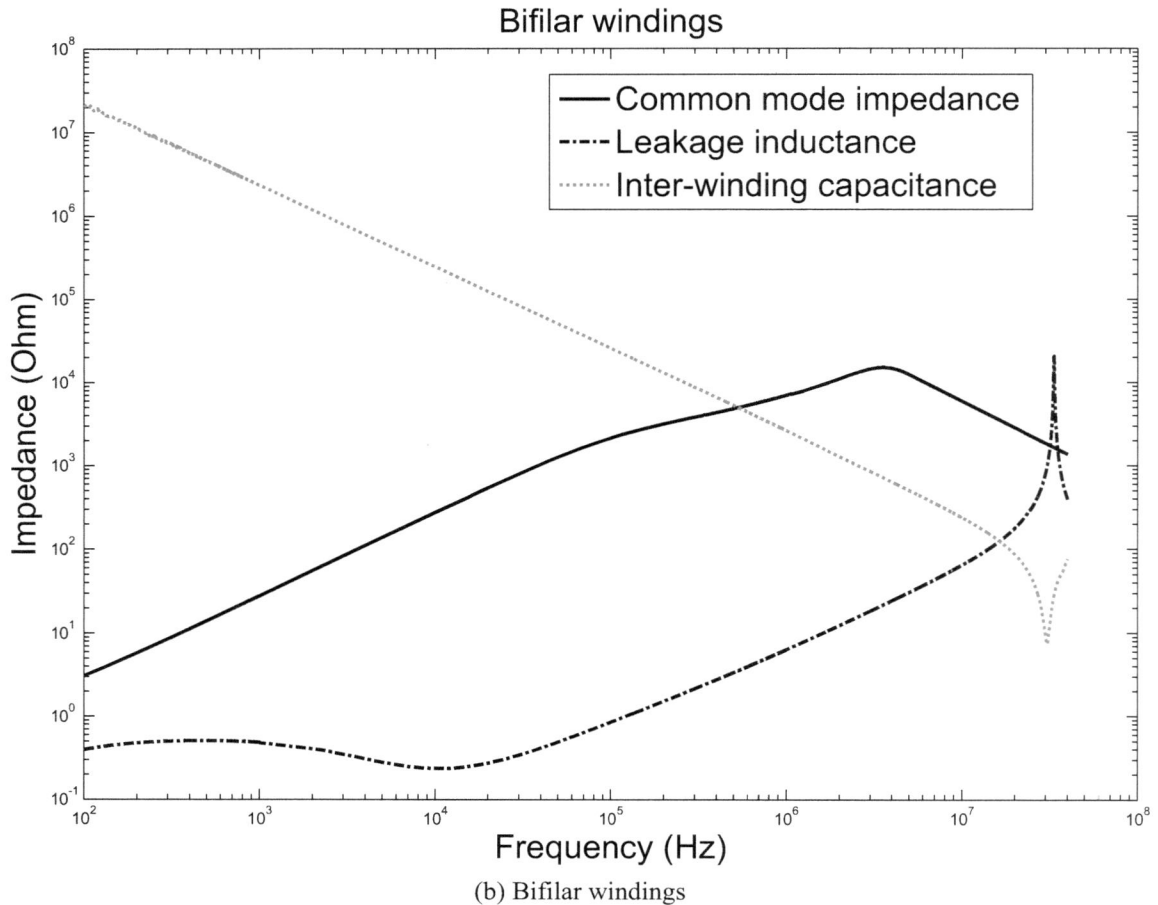

(b) Bifilar windings

Figure 36: Nanocrystalline common mode chokes impedances with different windings.

REFERENCES

[1]   Bose, Modern power electronics and AC drives, ISBN 0-13-016743-6, Prentice Hall.

[2]   Kempski A, Smolenski R, Kot E, Fedyczak Z. Active and passive series compensation of common mode voltage in adjustable speed drive system, Industry Applications Conference, 2004. 39th IAS Annual Meeting. Conference Record of the 2004 IEEE, 2004; 4: pp. 2665-2671.

[3]   Roc'h A, Bergsma H, Leferink FBJ, Zhao D, Polinder H, Ferreira JA. Design of an EMI Output Filter For Frequency Converters, in Proc. EMC Europe International Symposium on EMC, Barcelona, Spain, 2006.

[4]   Frank Leferink, Hans Bergsma, Braham Ferreira, Wim van Etten. High Performance EMI Filter for Frequency Converters, EMC Europe 2004, Eindhoven.

[5]   'Introduction to electromagnetic compatibility' Clayton R. Paul, ISBN 978-0-471-75500.

[6]   EMI Self-Learning Toolkit for Switched-Mode Power Supply, AUNP program, http://www.kmitl.ac.th/emc/emitoolkit.htm

[7]   Massarini A, Kazimierczuk MK. Self-capacitance of inductors. IEEE Trans Power Electron 1997; 12(4): 671-676.

[8]   Grandii G, Kazimiercmk Massarini A. Lumped. Parameter for models for single- and Multiple-layer Inductor, IEEE PESC '96.

[9]   Agilent technologies Impedance measurement handbook July 2006, www.agilent.com

[10]   Fairites, Technical Information, How to choose Ferrite Components for EMI suppression, www.fair-rite.com

[11]   Takanori T. Frequency dispersion of complex permeability in Mn-Zn and Ni-Zn spinel ferrites and their composite materials. J Appl Phys 2003; 93(5): 2789-2796.

[12]   Nave MJ, Technol Sverdrup, AL Huntsville. On modeling the common mode inductor Electromagnetic Compatibility, 1991. Symposium Record. IEEE 1991 International Symposium on, 12-16 Aug 1991; 452-457.

[13]   Margarita D. Takach, Peter O. Lauritzen: Survey of Magnetic Core Models, IEEE APEC Record, 1995; pp. 560-566.

[14]   Jiles DC. Atherton DL. Theory of ferromagnetic hysteresis. J Magn Magn Mater 1986; 61: 48-56.

[15]　NL Mi, Oruganti R, SX Chen. Modeling of hysteresis loops of ferrite cores excited by a transient magnetic field Magnetics, IEEE Trans 1998; 34(4): 1294-1296

[16]　Jiles DC. Frequency dependence of hysteresis curves in 'non-conducting' magnetic materials, Magnetics. IEEE Trans; 1993; 29(6): 3490-3492.

[17]　Izydorczyk J. Magnetics IEEE Trans 2006; 42(10): 3132–3134.

[18]　Saltelli T, Campolongo R. Sensitivity Analysis in Practice_ A Guide to Assessing Scientific Models, 2004.

[19]　Vlach, Singhal, Computer methods for circuit analysis and design, 1983, ISBN 978-0442011949.

[20]　Vacuumschmelze Gmbh&Co KG, Technical Information, Nanocrystalline material in common mode chokes.

[21]　Magnetics, Technical Information. A Critical Comparison of Ferrites with Other Magnetic Materials, www.mag-inc.com

[22]　Handbook of Magnetic Material, vol. 1, 2 and 6.ISBN: 0-444-51459-7.

[23]　Handbook of modern ferromagnetic material, Alex Goldman. ISBN: 0-412-14661-4.

[24]　Inductive components and appropriate magnetic materials for filters in SMPS-designs; Wanior, J, Industrial Technology, 2003 IEEE International Conference on Volume 2, 10-12 Dec. 2003, Vol. 2, pp.1178–1183.

[25]　Fe-M-B nanocrystalline soft magnetic materials: a review of soft magnetic material development, Kiyonori Suzuki, Properties and application of Nanocrystalline alloys form amorphous precursors.

[26]　Switched mode power supply handbook, Billings, Keith, H., ISBN 0-07-005330-8.

[27]　Electric Motors and Drives, Austin Hughes, ISBN-13: 978-0-7506-4718-2.

CHAPTER 2

Noise Source Impedance Measurement in SMPS

Vuttipon Tarateeraseth[1,*], Kye Yak See[2] and Flavio G. Canavero[3]

[1]*College of Data Storage Innovation, King Mongkut's Institute of Technology, Ladkrabang, Chalongkrung Rd., Ladkrabang, Bangkok, Thailand, 10520;* [2]*School of Electrical and Electronic Engineering, Nanyang Technological University, Block S1, Level B1C, Room 100, Nanyang Link, Singapore, 639798 and* [3]*Dipartimento di Elettronica, Politecnico di Torino, Torino, Corso Duca degli Abruzzi, 24 - 10129, Torino, Italy*

Abstract: An accurate measurement method to extract common mode (CM) and differential mode (DM) noise source impedances of a switched-mode power supply (SMPS) under operating condition is presented in this chapter. With a proper pre-measurement calibration process, the proposed method allows extraction of both the CM and the DM noise source impedances with very good accuracy. These noise source impedances come in handy to systematically design an electromagnetic interference filter for a SMPS with minimum hassle.

Keywords: Calibration process, CM noise source impedance, Common mode chokes, Common mode inductors, Conducted Electromagnetic Interferences, Conducted emissions, Coupling capacitors, Differential mode chokes, Differential mode inductors, Differential mode noise impedance, Direct clamping two-probe approach, DM noise source impedance, Equivalent series inductance (ESL), Equivalent series resistance (ESR), Hilbert transform, Insertion Loss (IL), Linearization, Load termination impedances, Mutual inductance, Noise coupling mechanism, Noise source impedance measurement, Noise source impedances, Noise termination impedances, Parasitic capacitances, Parasitic inductances, Power line impedances, Reflection coefficient, Reverse-recovery phenomena, RF instruments, RF isolation, Scattering parameters, Sensing probe, Small-signal perturbation, S-parameters, Switched-mode power supplies (SMPS), Vector Network Analyser (VNA), Y-capacitors (Y-caps), X-capacitors (X-caps).

STATE-OF-THE-ART OF NOISE SOURCE IMPEDANCE MEASUREMENT IN SMPS

Built-in power line electromagnetic interference (EMI) filters are parts of a switched-mode power supply (SMPS) to limit conducted EMI in the frequency range up to 30 MHz in order to comply with the international EMI regulatory requirements [1] - [2]. Unlike the filter designs for communications and microwave applications, where the source and termination impedances are well defined (usually specified at 50 Ω, the noise source impedance of a SMPS is far from 50 Ω [3], and is not readily available. On the other hand, in the standard conducted EMI measurement setup, the SMPS is powered through the line impedance stabilization network (LISN) whose impedance is well-defined [4]. One could think of estimating the noise source impedance of a SMPS using the datasheet or typical values, but the reliability of such estimates is somewhat questionable. In fact, the noise source impedance differs from the nominal SMPS (possibly provided by the manufacturer), due to converter topology, component parasitics, printed circuit board layout, *etc.* [5]. For example, the differential mode (DM) noise source impedance is strongly influenced by the reverse recovery phenomena of a diode rectifier [6], an equivalent series resistance (ESR) and an equivalent series inductance (ESL) of a bulk capacitor [7]. As for the common mode (CM) noise source impedance, the deciding components are the parasitic capacitance between the switching device and its heat-sink and the parasitic capacitance between the board and the chassis [8]. Also, the adequate models for SMPS are a hard task to derive because the complexity of the noise coupling mechanism. Thus, engineers prefer to resort to characterization measurements. A SMPS may not be easily modeled [9]. Hence, the design of an EMI filter without known noise source impedances can be a challenging task [10].

To illustrate why noise source and load termination impedances are the important parameters for an EMI filter design, the simple CL-configuration filter as shown in Fig. **1** (**b**) is analyzed. Generally, the EMI filter characteristics are defined by their insertion losses (IL) [1]. The IL is defined by the ratio of voltages appearing across the load before and after the filter insertion [11]. The IL is usually expressed in decibels (dB) as follows

$$IL = 20\log\left|\frac{V_o}{V_o'}\right| \tag{1}$$

*Address correspondence to Vuttipon Tarateeraseth: College of Data Storage Innovation, King Mongkut's Institute of Technology Ladkrabang, Chalongkrung Rd., Ladkrabang, Bangkok, Thailand, 10520; Tel: +66-80-5650072; E-mail: ktvuttip@kmitl.ac.th.

where

V_O = output voltage at the load without a filter [V]

V_O' = output voltage at the load after a filter insertion [V]

From Figs. **1** (**a**) and (**b**), the insertion loss of a simple CL-configuration filter can be determined by

$$IL = 20\log\left| s^2\left(\frac{LCZ_S}{Z_L+Z_S}\right) + s\left(\frac{L+CZ_SZ_L}{Z_L+Z_S}\right) + 1\right|$$ (2)

where

$s = j2\pi f$

Z_S = source impedance [Ω]

Z_L = load impedance [Ω]

C = filter capacitance [F]

L = filter inductance [H]

If the load termination impedance is fix at 50 Ω but the source impedance takes the values of 1 Ω, 50 Ω and 1 kΩ, the insertion loss can be determined according to (2) and plotted as shown in Fig. **1** (**c**).

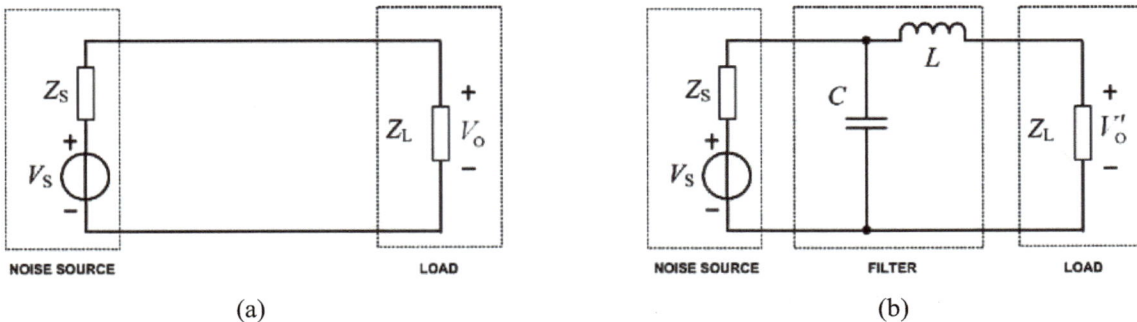

(a)   (b)

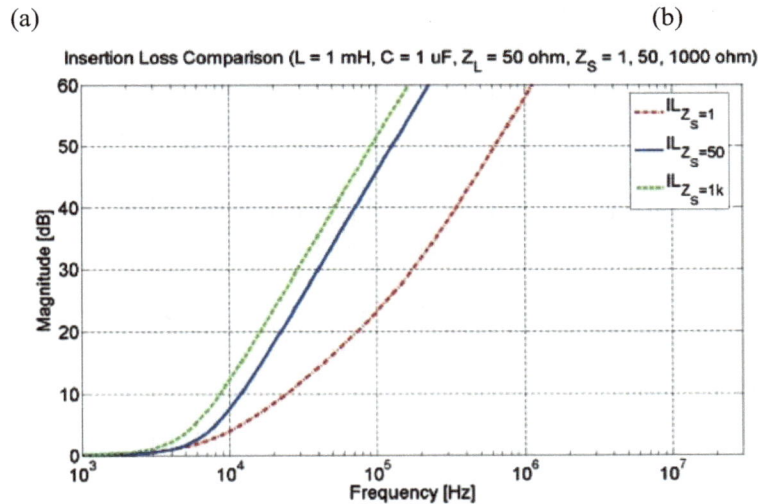

(c)

Figure 1: Effect of noise source and load termination impedances on the filter performance. (**a**) Direct connection of load to source; (**b**) filter insertion between source and load. Panels (**a**) and (**b**) are needed to clarify the IL definition. (**c**) Comparison of IL curves for different values of the source impedance, while the load is fixed at 50 Ohm.

From Fig. **1** (**c**), for example at 100 kHz, it can be seen that the insertion loss varies from about 20 dB to 50 dB. Normally, the EMI filter components are designed based on the insertion loss at a particular frequency [5]. Hence, designing an EMI filter for a SMPS by assuming a value of 50 Ω for the noise source and termination impedances, will lead to non-optimal EMI filter components. The need of information of the noise source impedances of a SMPS is already claimed in [12] - [15].

Some progress has been made to measure the DM and CM noise source impedances of a SMPS. In the first place, the resonance method was developed to estimate the noise source impedance of a SMPS by making a simplified assumption that the noise source is a simple Norton equivalent circuit made of a current source with parallel resistive and capacitive elements [16]. By terminating at the AC power input of the SMPS with a resonating inductor, the noise source impedance can be estimated [12]. However, the process to select and to tune the resonating inductor for resonance can be tedious and cumbersome. Also, when frequency increases, the parasitic effects of the non-ideal reactive components become significant and the circuit topology based on which the resonance method is developed is no longer valid. This simplistic approach provides only a very rough estimate of the noise source equivalent circuit model.

In the past, the insertion loss method was introduced to measure the DM and CM noise source impedances of a SMPS [17]. This method requires some prior conditions to be fulfilled. For example, the impedances of the inserted components must be much larger or smaller than the noise source impedances [18]. Hence, the accuracy deteriorates if these conditions are not met. Moreover, this approach is to measure only the magnitude of the noise source impedance. Although [17] suggests to reconstruct the phase by means of the classical Hilbert transform approach; in reality often, only magnitude of the noise source impedances is taken into consideration, in order to avoid complex mathematical manipulations.

Recently, a two-probe approach to measure the DM and CM noise source impedances of a SMPS was developed [19]. An injection probe, a sensing probe and some coupling capacitors are used in the measurement setup. In order to measure the DM and CM noise source impedances with reasonable accuracy, careful choices of the DM and CM chokes are necessary to provide very good RF isolation between the SMPS and the LISN. Moreover, special attention is needed to ascertain that the DM and the CM chokes are not saturated for a high power SMPS. Again, this method focuses on extracting the magnitude information of the noise source impedance only.

In view of the limitations of the previously discussed methods, a direct clamping two-probe approach is proposed. Unlike the former two-probe method [19], the proposed method, [20], uses direct clamp-on type current probes and therefore there is no direct electrical contact to the power line wires between the LISN and the SMPS. Hence, it eliminates the need of the coupling capacitors. Also, no isolating chokes are needed, making the measurement setup simple to implement. With a vector network analyzer as a measurement instrument, both the magnitude and the phase information can be extracted directly without further processing. The proposed method is also highly accurate as it has the capability to eliminate the error introduced by the measurement setup.

The assumption underlying our extraction procedure is, that the input impedance of the SMPS behaves linearly. This is reasonably true, since - according to [17] - the "on" state impedance prevails during operation, and the impedance probing is done by means of small-signal perturbations, thus allowing linearization [21].

This chapter is organized as follows. Section **2** provides the necessary background theory of the direct clamping two-probe measurement technique. Experimental validation of the proposed method is given in Section **3**. Section **4** describes the setups to measure the DM and CM noise source impedances of the SMPS under operating conditions. Finally, the conclusions are given in Section **5**.

$$
\begin{bmatrix} V_1 \\ 0 \\ -V_{a-a'} \end{bmatrix} = \begin{bmatrix} 50\Omega + Z_{p1} & 0 & -j\omega M_1 \\ 0 & 50\Omega + Z_{p2} & +j\omega M_2 \\ -j\omega M_1 & +j\omega M_2 & r_w + j\omega L_w \end{bmatrix} \begin{bmatrix} I_1 \\ I_2 \\ I_w \end{bmatrix} \qquad (3)
$$

THEORY OF THE DIRECT CLAMPING TWO-PROBE MEASUREMENT

Two-probe measurement technique was first applied to measure the impedance of the equipment under test (EUT), *e.g.* operating small AC motor, fluorescent light, *etc.*, within frequency range from 20 kHz to 30 MHz [22]. Then, the power line impedance was measured within frequency range from 10 kHz to 30 MHz [23] and from 20 kHz to 30 MHz [24]. Later, the frequency range of the power line impedance measurement is extended up to 500 MHz [25]. It is worth to note that the measured starting frequency is about 10 kHz because, prior to 1987, most RFI instruments started to measure from 10 kHz, and it was considered as lower frequency [2]. For the stopped frequency, 30 MHz is normally used since it is the maximum frequency range of conducted emissions [1] - [2], [4]. With the same concept, the characterizations of the DM and CM noise source impedances of the SMPS can be extracted as proposed in [19], [20].

The basic concept of the direct clamping two-probe method to measure any unknown impedance is illustrated in Fig. **2**. It consists of an injection current probe, a detection current probe and a Vector Network Analyzer (VNA). Port 1 of the VNA generates an AC signal into the closed loop through the injection probe and the resulting signal current in the loop is measured at port 2 of the VNA through the detection probe.

Fig. **3** shows the complete equivalent circuit of the measurement setup shown in Fig. **2** where V_1 is the output signal source voltage of port 1 connected to the injection probe and V_{p2} is the resultant signal voltage measured at port 2 with the detection probe. The output impedance of port 1 and the input impedance of port 2 of the VNA are both 50 Ω. L_1 and L_2 are the primary inductances of the injection and the detection probes, respectively. L_w and r_w are the inductance and the resistance of the wiring connection that formed the circuit loop, respectively. M_1 is the mutual inductance between the injection probe and the circuit loop and M_2 is the mutual inductance between the detection probe and the circuit loop. Z_{p1} and Z_{p2} are the input impedances of the injection and detection probes, respectively. The excitation signal source V_1 induces a signal current I_w in the circuit loop through the injection probe.

From Fig. **3**, we can derive the equations describing the three circuits as shown in Eq. (3). Eliminating I_1 and I_2 from (3), we obtain

$$V_{M1} = V_{a-a'} + \left(Z_{M1} + Z_{M2} + r_w + j\omega L_w \right) I_w \tag{4}$$

where $Z_{M1} = \dfrac{\left(\omega M_1 \right)^2}{50\,\Omega + Z_{p1}}$, $Z_{M2} = \dfrac{\left(\omega M_2 \right)^2}{50\,\Omega + Z_{p2}}$ and $V_{M1} = V_1 \left(\dfrac{j\omega M_1}{50\,\Omega + Z_{p1}} \right)$

According to expression (4), the injection probe can be reflected in the closed circuit loop as an equivalent current-controlled voltage source V_{M1} in series with a reflected impedance Z_{M1} and the detection probe can be reflected in the same loop as another impedance Z_{M2}, as shown in Fig. **4**. For frequencies below 30 MHz, the dimension of the coupling circuit loop is electrically small as compared to the wavelengths concerned. Therefore, the current distribution in the coupling circuit is uniform throughout the loop, and V_{M1} can be rewritten as

$$\begin{aligned} V_{M1} &= \left(Z_{M1} + Z_{M2} + r_w + j\omega L_w + Z_x \right) I_w \\ &= \left(Z_{setup} + Z_x \right) I_w \end{aligned} \tag{5}$$

The equivalent circuit seen at $a - a'$ by the unknown impedance Z_x can be substituted by an equivalent current-controlled voltage source V_{M1} in series with an impedance due to the measurement setup Z_{setup}. From (5), Z_x can be determined by

$$Z_x = \frac{V_{M1}}{I_w} - Z_{setup} \tag{6}$$

According to the detection probe loop of Fig. **4**, the current I_w measured by the detection probe is

$$I_w = \frac{V_{p2}}{Z_{T2}} \tag{7}$$

where V_{p2} is the signal voltage measured at port 2 of the VNA and Z_{T2} is the calibrated transfer impedance of the detection probe provided by the probe manufacturer. Substituting V_{M1} and (7) into (6) yields

$$Z_x = \left(\frac{j\omega M_1 Z_{T2}}{50\ \Omega + Z_{p1}} \right) \left(\frac{V_1}{V_{p2}} \right) - Z_{setup} \tag{8}$$

The excitation source V_1 of port 1 of the VNA and the resultant voltage at the injection probe V_{p1} are related by

$$V_1 = \left(\frac{50\ \Omega + Z_{p1}}{Z_{p1}} \right) V_{p1} \tag{9}$$

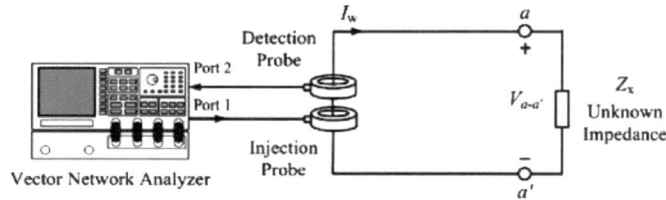

Figure 2: Conceptual direct clamping two-probe measurement.

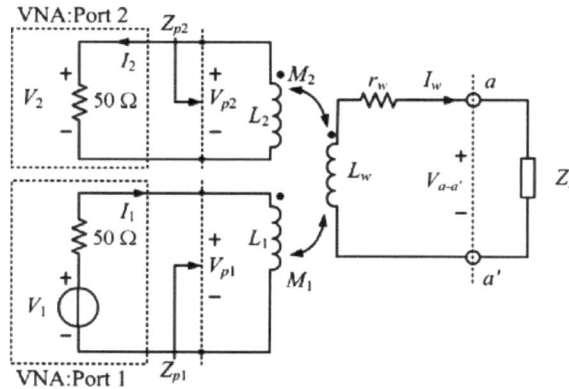

Figure 3: Equivalent circuit of the two-probe measurement setup of Fig. **2**.

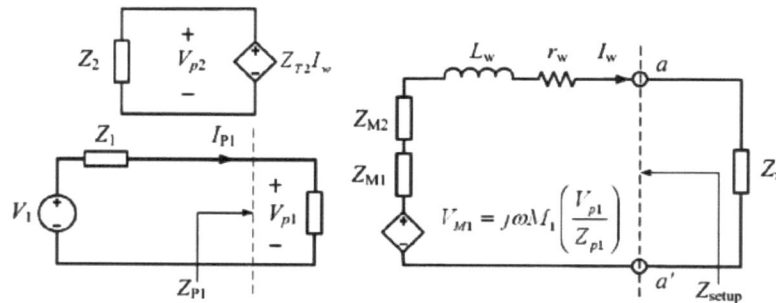

Figure 4: Final equivalent circuit of the setup connected to the unknown impedance.

Substituting (9) into (8), the unknown impedance can finally be expressed as

$$Z_x = K\left(\frac{V_{p1}}{V_{p2}}\right) - Z_{setup} \tag{10}$$

where $K = \left(\dfrac{j\omega M_1 Z_{T2}}{Z_{p1}}\right)$, which is a frequency dependent coefficient. The ratio V_{p1}/V_{p2} can be obtained through the S-parameters measurement using the VNA. From Fig. **3**, the resultant signal voltage source can be defined by $V_{p1} = (S_{11}+1)V_{p1}^+$ and the resulting measured voltage can be defined by $V_{p2} = S_{21}V_{p1}^+$ where V_{p1}^+ is the amplitude of the voltage wave incident on port 1 of the VNA [25] - [26]. As a result, the ratio of the two probe voltages is given by

$$\frac{V_{p1}}{V_{p2}} = \frac{S_{11}+1}{S_{21}} \tag{11}$$

where

S_{11} = the measured reflection coefficient at port 1

S_{21} = the measured forward transmission coefficient at port 2

The coefficient K and the setup impedance Z_{setup} can be obtained by the following steps. Firstly, measure V_{p1}/V_{p2} by replacing impedance Z_x with a known precision standard resistor R_{std}. As a rule of thumb, the resistance of R_{std} should be chosen somewhere in a middle range of the unknown impedance to be measured. Then, measure V_{p1}/V_{p2} again by short-circuiting $a-a'$. With these two measurements and (10), two equations (12) - (13) with two unknowns K and Z_{setup} result. Hence, K and Z_{setup} can be obtained by solving (12) and (13). Once K and Z_{setup} are found, the two-probe setup is ready to measure any unknown impedance using (10).

$$R_{std} = K\left(\frac{V_{p1}}{V_{p2}}\right)\Bigg|_{Z_x = R_{std}} - Z_{setup} \tag{12}$$

$$0 = K\left(\frac{V_{p1}}{V_{p2}}\right)\Bigg|_{Z_x = short} - Z_{setup} \tag{13}$$

It is ought to be noted that, for the sake of clarity, Fig. **2** is simplified and does not contain the LISN powering the active device under test (the SMPS, in our case). The LISN impedance should be considered as a part of Z_{setup}, without limitations. An additional remark is that the injected signal of the VNA must be much larger than the background noise generated by the device under test in the frequency range of interest, so that the background noise does not alter the Z_x value, superimposing on the measured quantities. For most low and medium power active systems, such a condition can usually be met. However, if the active system is characterized by very high power and generates significant background noise, one could add a power amplifier at the output of port 1 of the VNA to increase the power of the injected signal, so that the above condition could be fulfilled. Moreover, a pre-measurement calibration process not properly set jeopardizes the accuracy of the proposed method.

EXPERIMENTAL VALIDATION

In the experimental validation that follows, the Solar 9144-1N current probe (10 kHz - 100 MHz) and the Schaffner CPS-8455 current probe (10 kHz - 1000 MHz) are chosen as the injection and the detection current probes, respectively. The R&S ZVB8 VNA (300 kHz - 8 GHz) is selected for the S-parameter measurement.

Practically, a DM noise source impedance of a SMPS ranges from several ohms to several tens of ohms and a CM noise source impedance is capacitive in nature and is in the range of several $k\Omega$ [5], [12], [15]. In the validation, a precision resistor R_{std} ($620\,\Omega \pm 1\%$) is chosen, as it is somewhere in the middle of the range of unknown impedance to be measured (tens of Ω to a few $k\Omega$). Based on the procedure described in Section 2, K and Z_{setup} are determined accordingly. Once K and Z_{setup} are found, a few resistors of known values (2.2Ω, 12Ω, 24Ω, 100Ω, 470Ω,

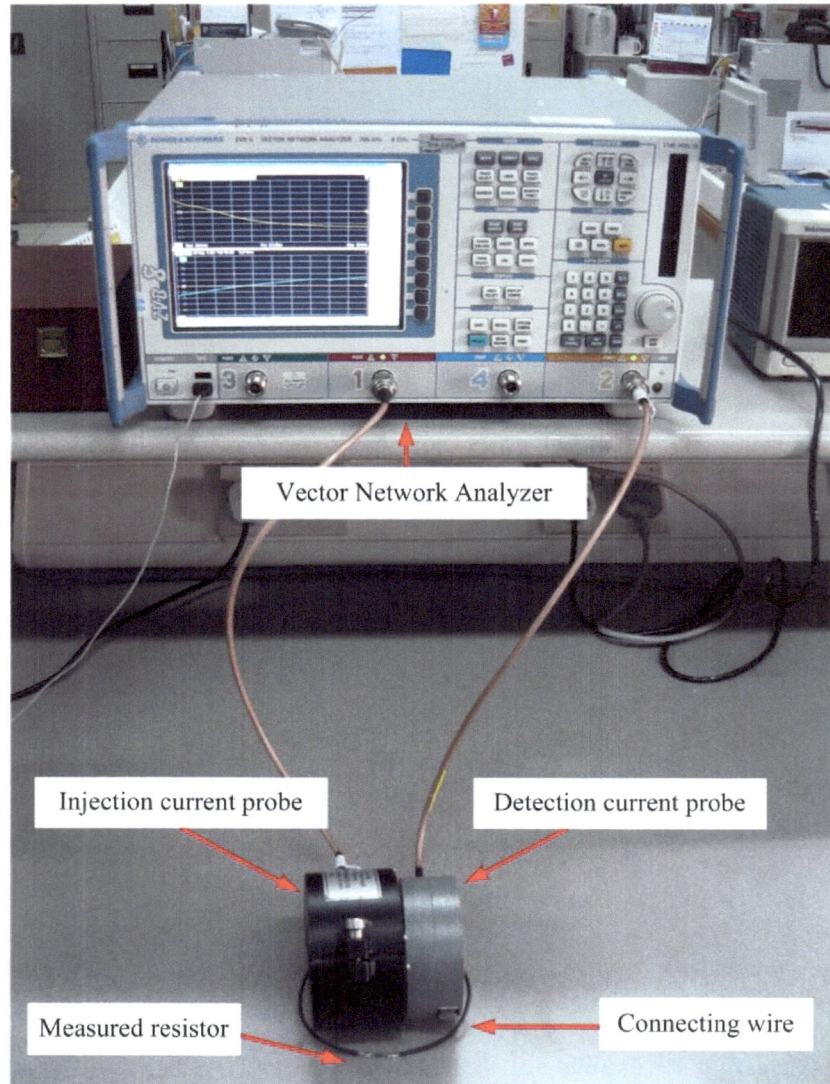

Figure 5: Photograph of measurement setup of selected resistors using the direct clamping two-probe approach.

$820\,\Omega$, $1.8\,k\Omega$ and $3.3\,k\Omega$) are treated as the unknown impedances and measured by the direct clamping two-probe setup as shown in Fig. 5. The wire loop to the resistor-under-measurement is made as small as possible to avoid any loop resonance below 30 MHz. Even by making the loop very small, there is still a finite impedance due to the measurement setup $\left(Z_{setup}\right)$. Z_{setup} comprises of the effects of the injection and the detection probes, the wire connection to the resistor and the coaxial cable between the current probes and the VNA. The ability to measure Z_{setup} and to subtract it from the two-probe measurement eliminates the error due to the setup and provides highly accurate measurement results. The measurement frequency range is from 300 kHz to 30 MHz. As shown in Figs. **6** (a) and (b), the magnitude and the phase of the so called "unknown resistors" are measured by the proposed method. The measured results are in close agreement with the stated resistance values of the resistors.

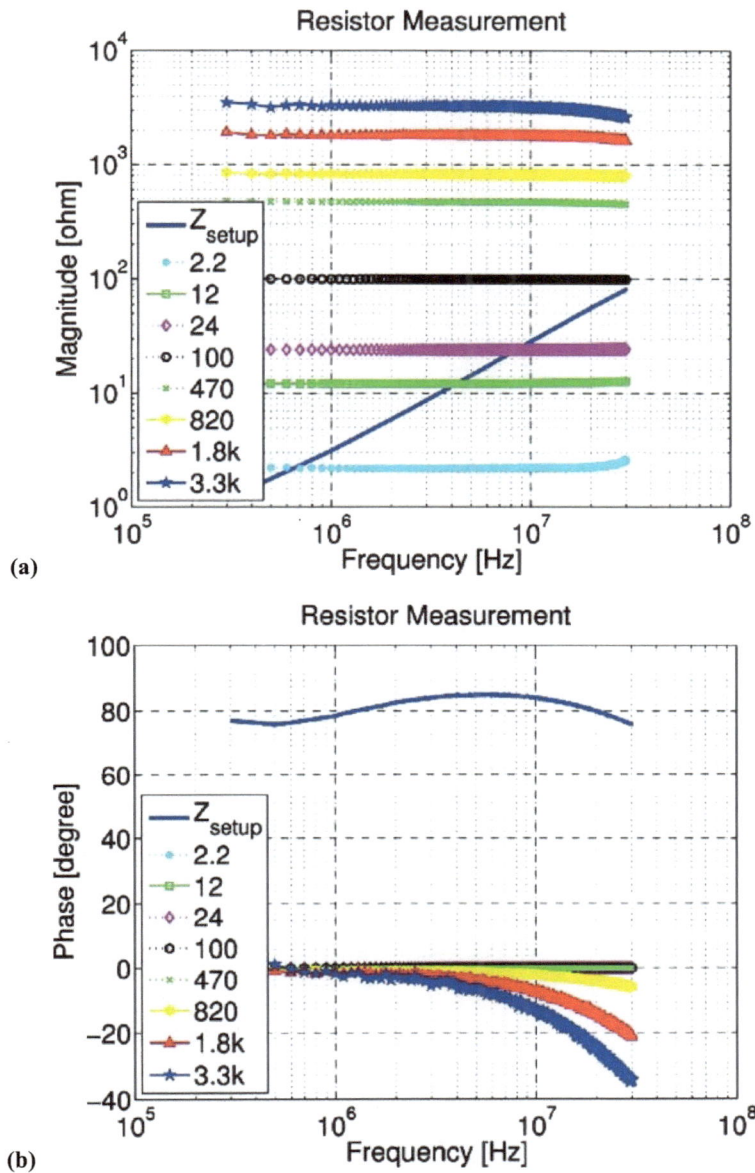

Figure 6: Measured results of selected resistors using the proposed two-probe approach. (**a**) Magnitude; (**b**) phase.

For large-resistance resistors such as 1.8 kΩ and 3.3 kΩ, the roll-off at higher frequency is expected due to the parasitic capacitance that is inherent to large-resistance resistors, but the parasitic effect is negligible for small resistance resistors. In Fig. **6**, Z_{setup} is also plotted to show its relative magnitude and phase with respect to the measured resistances. It shows that Z_{setup} is predominantly inductive and can be as high as 100 Ω at 30 MHz. Hence, for small-resistance resistors, whose values are comparable to Z_{setup}, the error contributed from Z_{setup} can be very large, if it is not subtracted from the measurement.

For further validation purposes, the DM as well as the CM output impedances of a LISN (Electro-Metrics MIL 5-25/2) are measured using the proposed method and the HP4396B impedance analyzer (100 kHz - 1.8 GHz). The measured DM impedance $\left(Z_{LISN,DM}\right)$ and the measured CM impedance $\left(Z_{LISN,CM}\right)$ of the LISN using both methods are compared. The experimental setup to measure the DM and CM output impedances of LISN using impedance analyzer is shown in Figs. **7 (a)** and (**b**), respectively.

By using the two-probe method, the LISN can be measured with the AC power applied. However, the measurement using the impedance analyzer can only be made with no AC power applied to the LISN to prevent damage to the measuring equipment. For the two-probe method, AC power is applied to the input of the LISN and one or two 1 μF "X class" capacitors are connected at the output of the LISN to implement an AC short circuit. It should be noted that because the impedance of 1 μF capacitor is very low at measured frequency range, its impedance is not taken into account. A 1 μF capacitor is connected between line and neutral wires for DM measurement as shown in Fig. **8** (**a**). For CM measurement, two 1 μF capacitors are needed, one connected between line and ground and another connected between neutral and ground as shown in Fig. **8** (**b**).

For the DM output impedance measurement, the line wire is treated as one single outgoing conductor and the neutral wire is treated as the returning conductor as shown in Figs. **8**(**a**) and **9**(**a**).

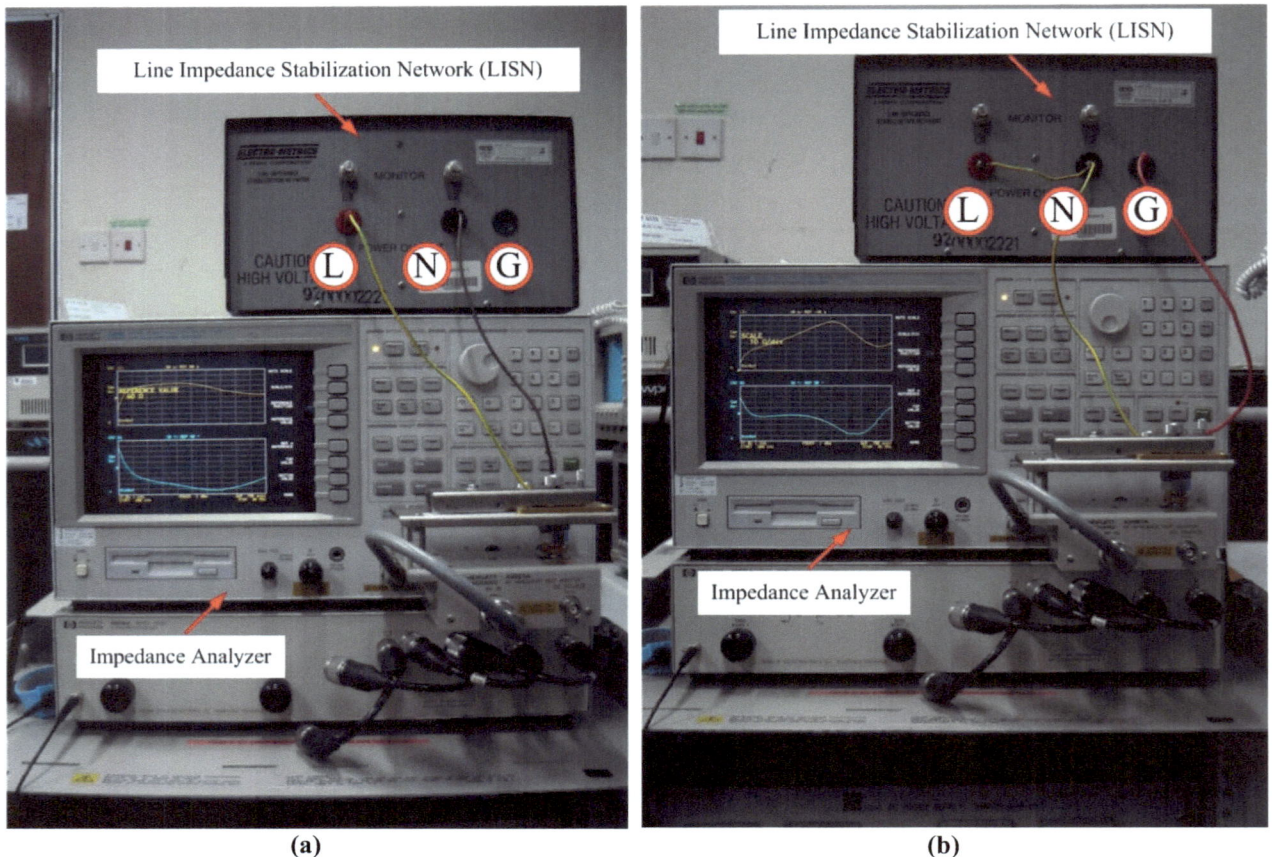

Figure 7: Photograph of impedance measurement setup of the LISN using an impedance analyzer. (**a**) DM; (**b**) CM.

In the case of CM measurement, the line and the neutral wires are treated as one single outgoing conductor, and the safety ground wire is treated as the returning conductor as shown in Figs. **8** (**b**) and **9** (**b**). The length of the connecting wire between the LISN and capacitor is chosen to be as short as possible to eliminate the parasitic inductance of the connecting wires. However, since the connecting wire is different from the case of the simple resistor measurements, the $Z_{setup,DM}$, $Z_{setup,CM}$ and the frequency dependent coefficient (K_{DM} and K_{CM}) need to be re-calibrated.

For DM impedance measurement, the injection and detection probes are clamped on the connecting line wire only as shown in Fig. **8** (**a**). The $Z_{setup,DM}$ and K_{DM} can be obtained by measuring V_{p1}/V_{p2} when both LISN and "X class" capacitors are removed, and short-circuiting the line and neutral wires at both ends. Again, we need to measure V_{p1}/V_{p2} by connecting the precision standard resistor R_{std} (620 $\Omega \pm 1\%$) at one end.

For the CM impedance measurement, since the line and the neutral wires are treated as one single outgoing conductor, both current probes are clamped onto the line and the neutral wires, as shown in Fig. **8 (b)**. The $Z_{setup,CM}$ and K_{CM} can be obtained by removing both LISN and two "X class" capacitors and measuring V_{p1}/V_{p2} when the line, neutral and ground wires are shorted at both ends. Then, we need to measure V_{p1}/V_{p2} by connecting the precision standard resistor R_{std} $(620\ \Omega \pm 1\%)$ between line-neutral and ground at one end.

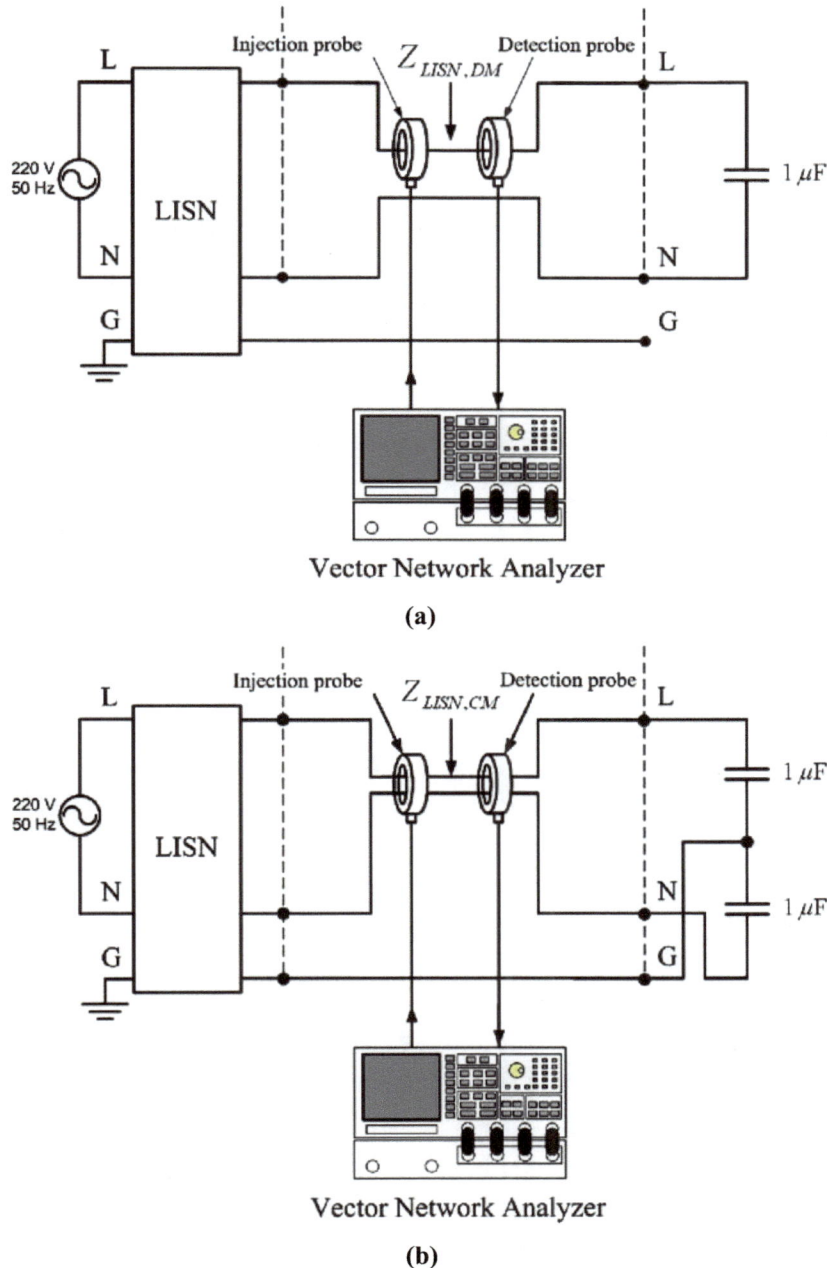

(a)

(b)

Figure 8: Impedance measurement setup of the LISN using direct clamping two-probe approach. **(a)** DM; **(b)** CM.

Substituting those measurement results into equations (12) and (13), the $Z_{setup,DM}$, $Z_{setup,CM}$ and the frequency dependent coefficients K_{DM} and K_{CM} can be obtained. Since the LISN schematics and component values are provided by the manufacturers or standards, the DM and CM impedances of the LISN can be readily calculated.

For comparison purposes, the simulated DM and CM impedances of the LISN using the datasheet provided by manufacturer are also plotted as shown in Figs. **10 (a)** - **(b)** and Figs. **11 (a)** - **(b)**, respectively.

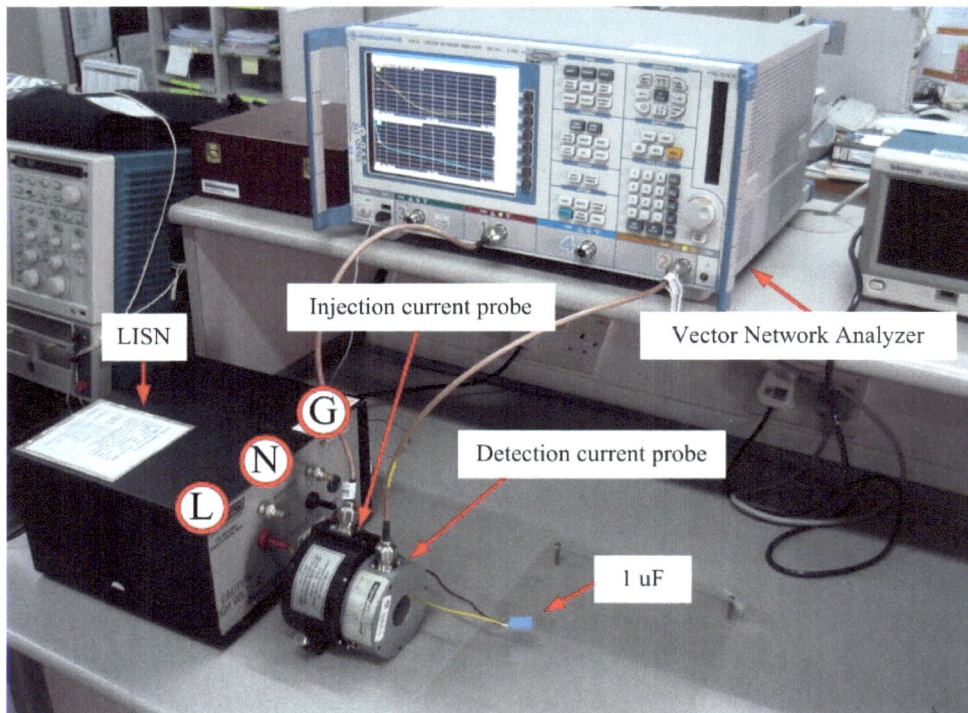

(a)

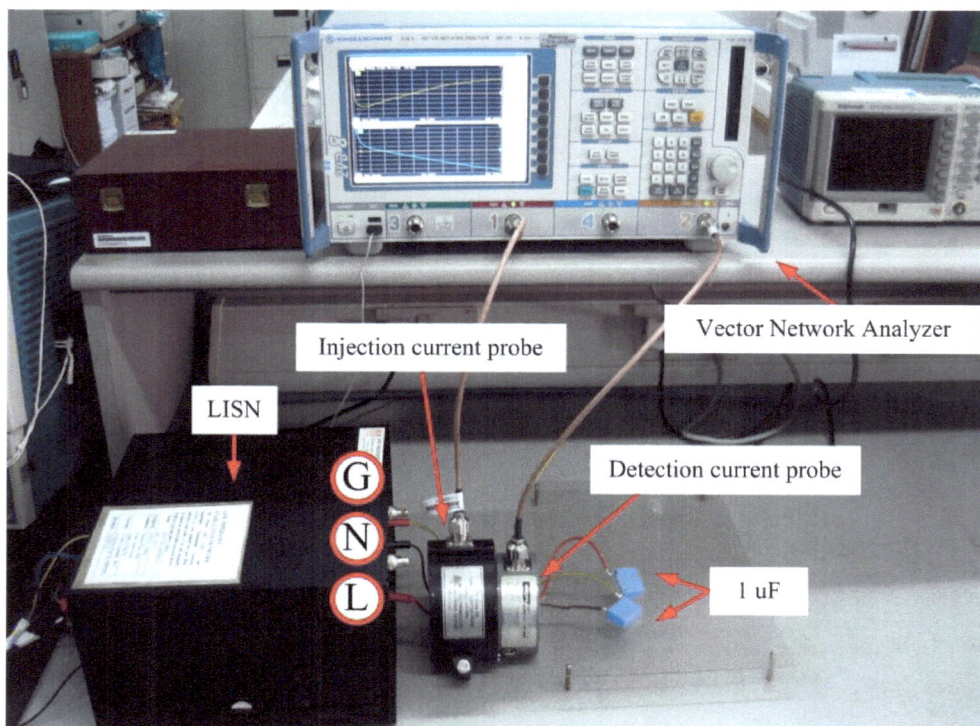

(b)

Figure 9: Photograph of LISN impedance measurement setup using direct clamping two-probe approach. (**a**) DM; (**b**) CM.

The comparisons of simulated output impedance of LISN ($Z_{LISN,DM(simul.)}$ and $Z_{LISN,CM(simul.)}$), measured output impedance of LISN using the direct clamping two-probe approach ($Z_{LISN,DM(2probes)}$ and $Z_{LISN,CM(2probes)}$), and using the impedance analyzer ($Z_{LISN,DM(IA)}$

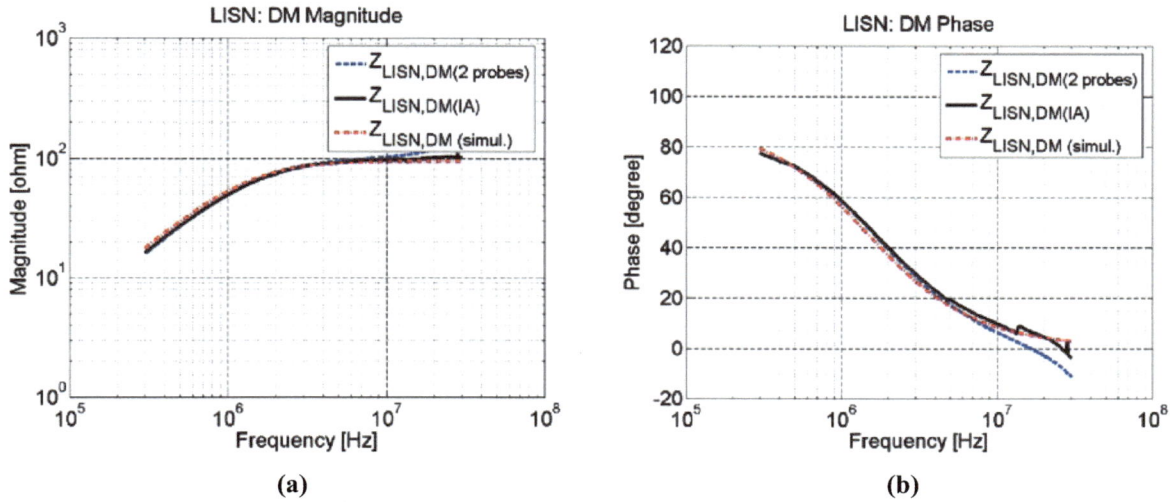

Figure 10: Comparison of measured results for LISN. (**a**) DM magnitude; (**b**) DM phase.

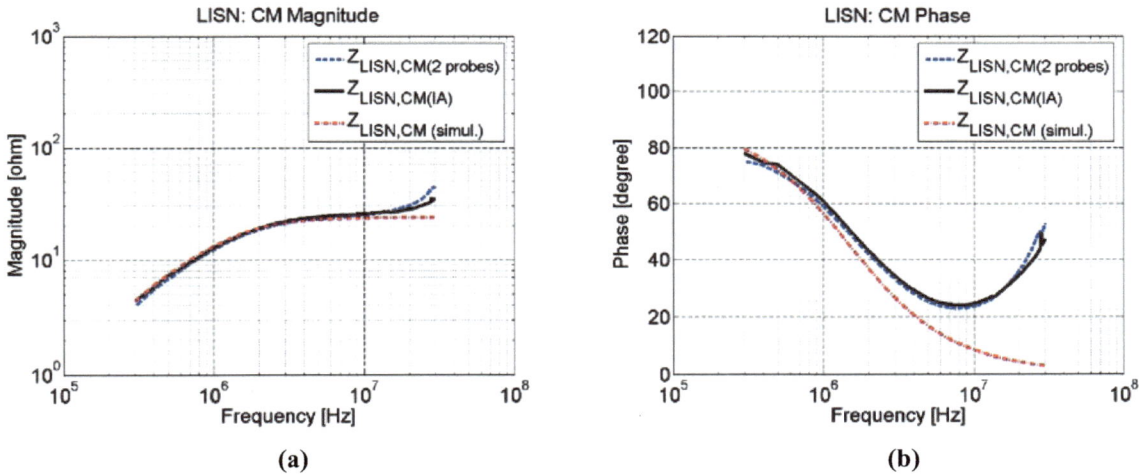

Figure 11: Comparison of measured results for LISN. (**a**) CM magnitude; (**b**) CM phase.

and $Z_{LISN,CM(IA)}$) are given in Figs. **10** and **11**, respectively. Again, close agreement among the simulated results and the two measurement methods is demonstrated.

MEASUREMENT OF NOISE SOURCE IMPEDANCES OF THE SMPS

The measurement setups to extract the DM noise source impedance $\left(Z_{SMPS,DM} \right)$ and the CM noise source impedance $\left(Z_{SMPS,CM} \right)$ of a SMPS are shown in Figs. **12** (**a**) - **13** (**a**) and Figs. **12** (**b**) - **13** (**b**), respectively. The technical specifications of the SMPS are: VTM22WB, 15 W, +12 V$_{dc}$/0.75 A, -12 V$_{dc}$/0.5 A. The SMPS is powered through the MIL 5-25/2 LISN to ensure stable and repeatable AC mains impedance. A resistive load is connected at the output of the SMPS for loading purposes. The DM impedance $\left(Z_{LISN,DM} \right)$ and the CM impedance $\left(Z_{LISN,CM} \right)$ of the LISN have been measured earlier in Section 3 and presented in Fig. **10** and **11**, respectively.

(a)

(b)

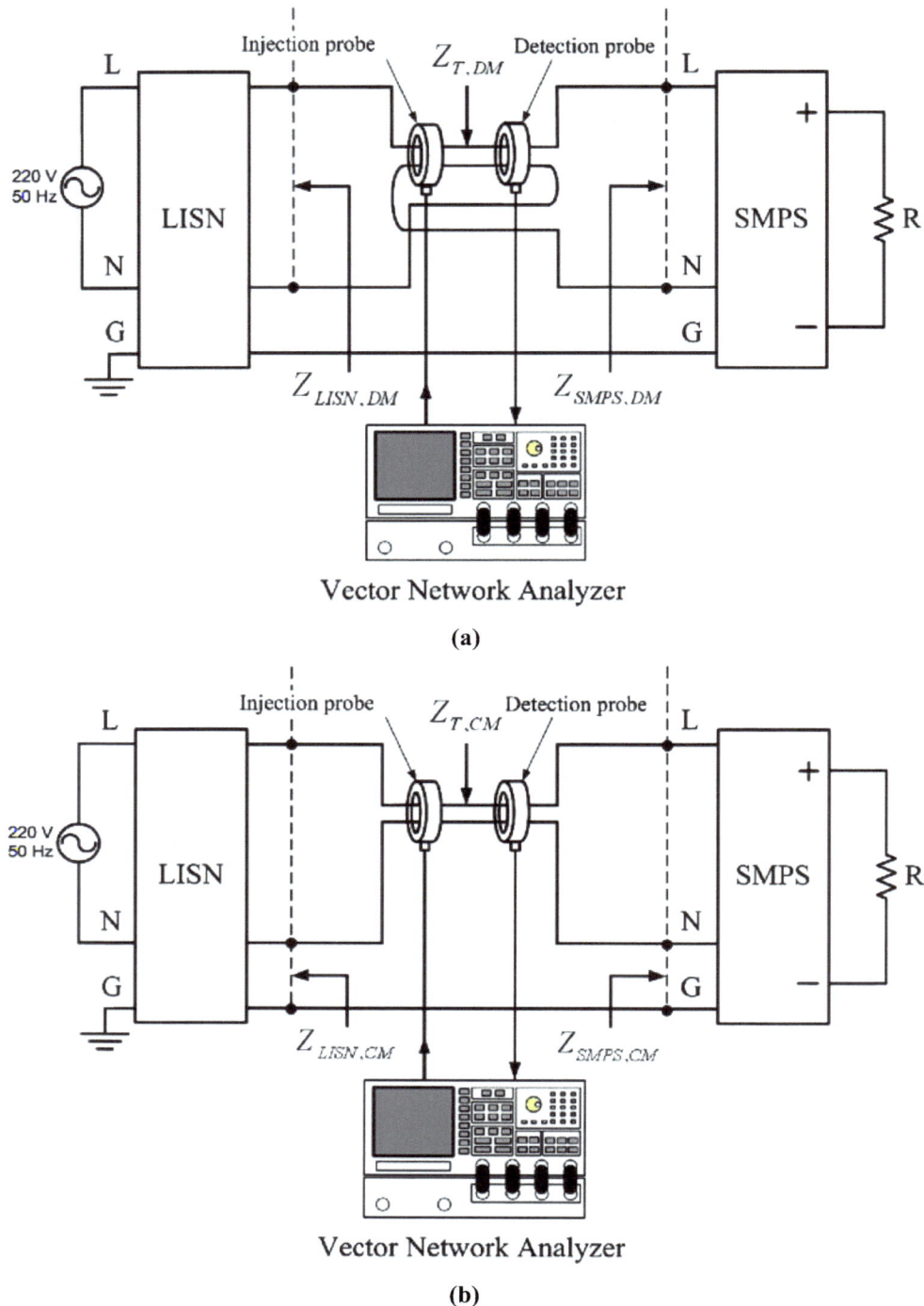

Figure 12: Noise source impedance measurement setup of the SMPS. (**a**) DM; (**b**) CM.

According to the measurement procedure of Section 2, we need first to extract the DM and the CM setup impedances, $Z_{setup,DM}$ and $Z_{setup,CM}$, and the frequency dependent coefficients K_{DM} and K_{CM} of the measurement setup. For DM impedance measurement calibration process, the injection and detection probes are clamped on the connecting line and neutral wires as shown in Fig. **12** (**a**) and **13** (**a**). The $Z_{setup,DM}$ and K_{DM} can be obtained by measuring V_{p1}/V_{p2} when both LISN and SMPS are removed and short-circuiting the line and neutral wires at both ends. Again, we measure V_{p1}/V_{p2} by connecting the precision standard resistor R_{std} (620 $\Omega \pm 1\%$) at one end.

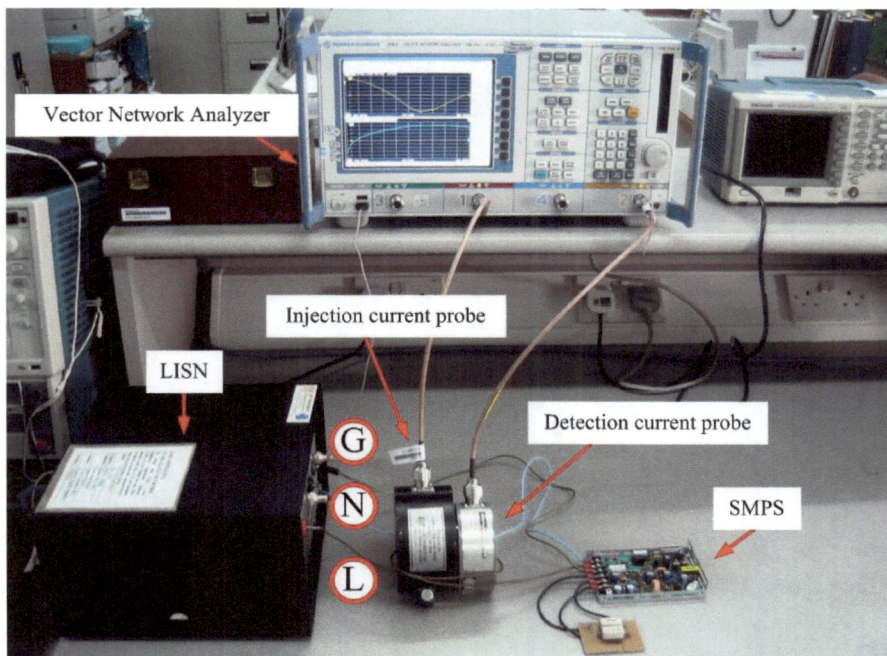

(a)

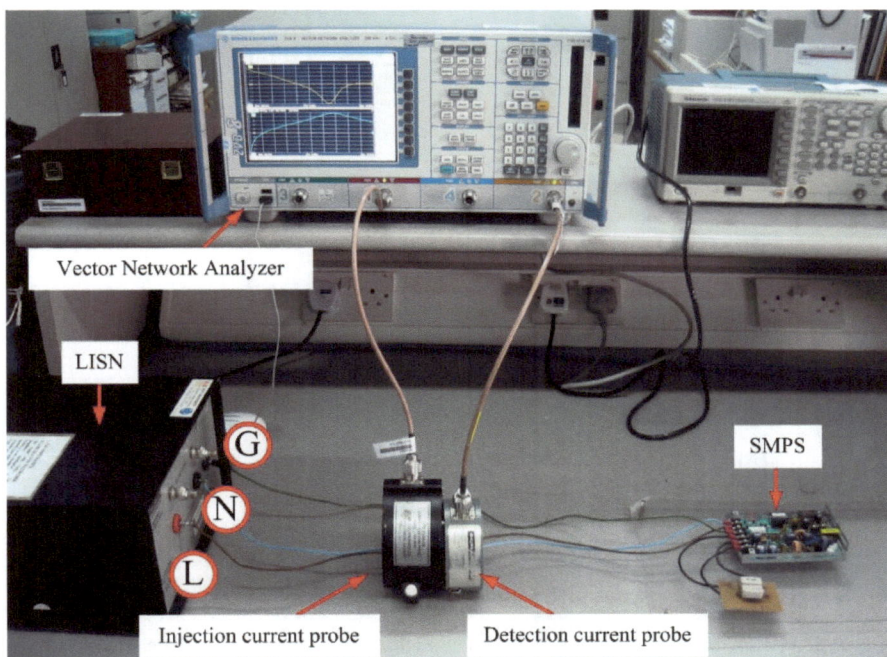

(b)

Figure 13: Photograph of noise source impedance measurement setup of the SMPS. (**a**) DM; (**b**) CM.

For the CM impedance measurement calibration process, both current probes are clamped onto the line and the neutral wires, as shown in Fig. **12** (**b**) and **13** (**b**). The $Z_{setup,CM}$ and K_{CM} can be obtained by removing both LISN and SMPS and measuring V_{p1}/V_{p2} when the line, neutral and ground wires are shorted at both ends. Then, we measure V_{p1}/V_{p2} by connecting the precision standard resistor R_{std} (620 Ω ± 1%) between line-neutral and ground at one end.

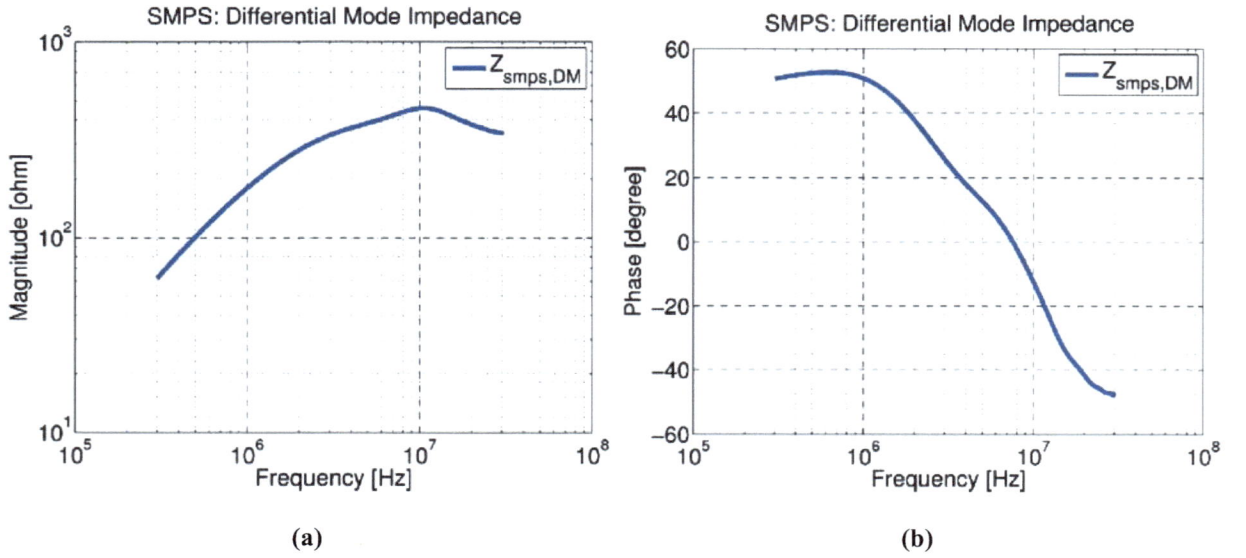

(a)   (b)

Figure 14: Noise source impedance measurement. (**a**) DM Magnitude; (**b**) DM Phase.

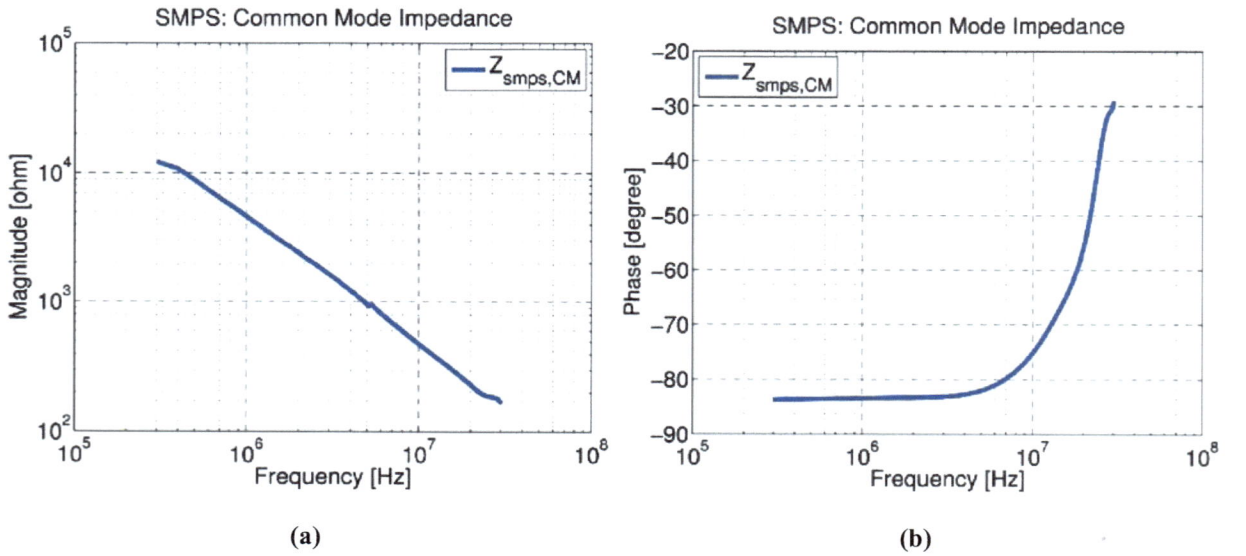

(a)   (b)

Figure 15: Noise source impedance measurement. (**a**) CM Magnitude; (**b**) CM Phase.

Substituting those measurement results into equations (12) and (13), the $Z_{setup,DM}$, $Z_{setup,CM}$ and the frequency dependent coefficients K_{DM} and K_{CM} can be obtained. The transmission line effect of the wire connection can be ignored as the length of connecting wires from the LISN to the SMPS is 70 cm, which is much shorter than the wavelength of the highest frequency of interest (30 MHz). The measured impedances are the total impedances in the circuit loops and we designate $Z_{T,DM}$ and $Z_{T,CM}$ as the total DM and CM impedances of the circuit loops connecting the SMPS and LISN under AC powered up operating conditions. As a result, $Z_{T,DM}$ and $Z_{T,CM}$ are defined by

$$Z_{T,DM} = Z_{LISN,DM} + Z_{SMPS,DM} + Z_{setup,DM} \qquad (14)$$

$$Z_{T,CM} = Z_{LISN,CM} + Z_{SMPS,CM} + Z_{setup,CM} \qquad (15)$$

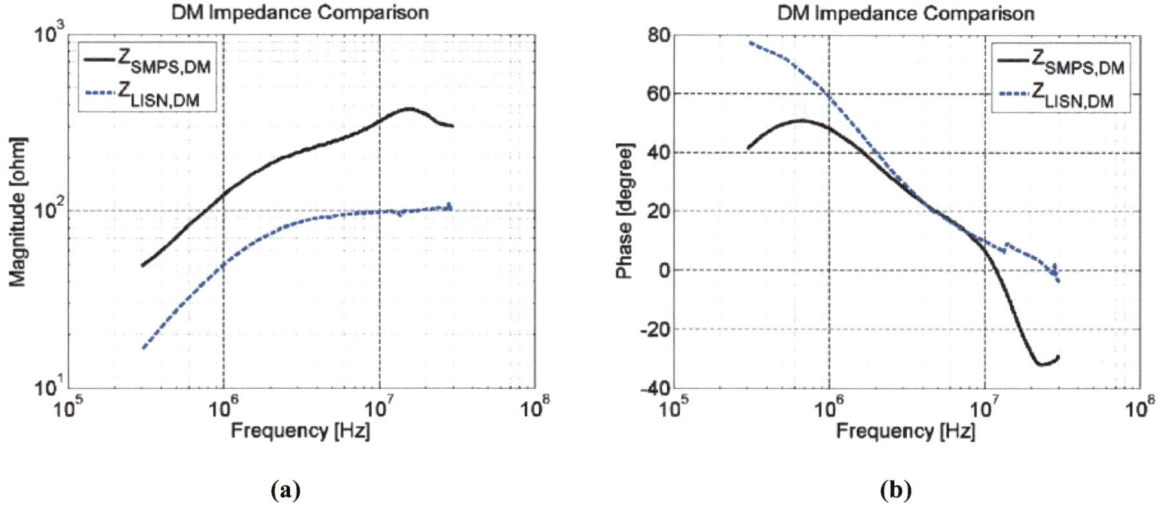

Figure 16: Comparison of LISN and SMPS impedances. (**a**) DM Magnitude; (**b**) DM Phase.

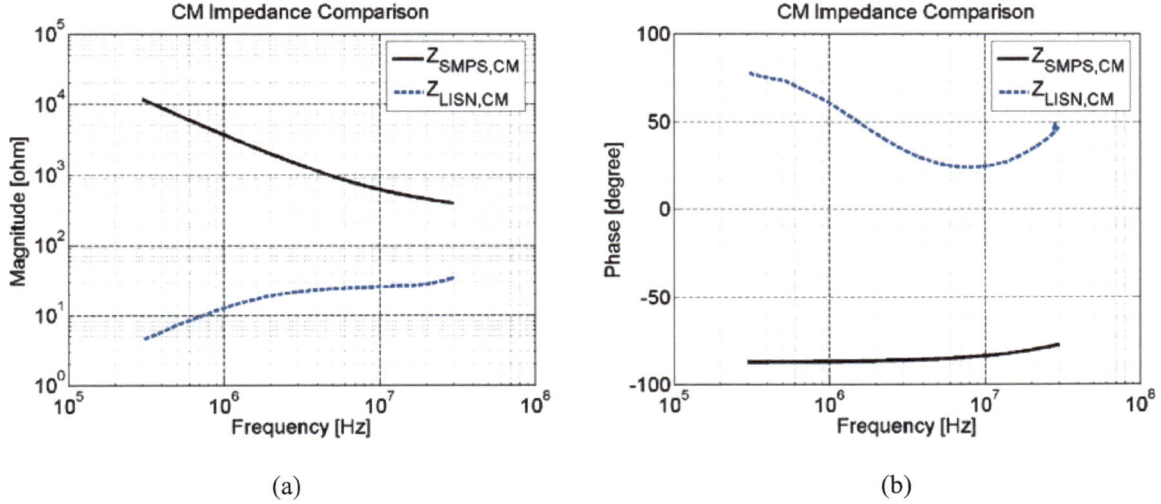

Figure 17: Comparison of LISN and SMPS impedances. (**a**) CM Magnitude; (**b**) CM Phase.

where

$$Z_{LISN,DM} = \text{DM output impedance of the LISN } [\Omega]$$

$$Z_{LISN,CM} = \text{CM output impedance of the LISN } [\Omega]$$

$$Z_{SMPS,DM} = \text{DM input impedance of the SMPS } [\Omega]$$

$$Z_{SMPS,CM} = \text{CM input impedance of the SMPS } [\Omega]$$

$$Z_{setup,DM} = \text{DM impedance due to the measurement setup } [\Omega]$$

$$Z_{setup,CM} = \text{CM impedance due to the measurement setup } [\Omega]$$

With known $Z_{LISN,DM}$, $Z_{LISN,CM}$, $Z_{setup,DM}$ and $Z_{setup,CM}$, once Z_T is measured, the DM and CM noise source impedances of the SMPS can be evaluated easily using (16) and (17), respectively.

$$Z_{SMPS,DM} = Z_{T,DM} - Z_{LISN,DM} - Z_{setup,DM} \tag{16}$$

$$Z_{SMPS,CM} = Z_{T,CM} - Z_{LISN,CM} - Z_{setup,CM} \tag{17}$$

Figs. **14** (**a**) and (**b**) show the magnitude and the phase of the extracted DM noise source impedance $\left(Z_{SMPS,DM}\right)$ in the frequency range from 300 kHz to 30 MHz. In general, the DM noise source impedance is dominated by the series inductive and resistive components at low frequencies; above 10 MHz, the effect of the diode junction capacitance of the full-wave rectifier begins to play a role. Figs. **15** (**a**) and (**b**) show the magnitude and the phase of the extracted CM noise source impedance $\left(Z_{SMPS,CM}\right)$. The CM noise source impedance is dominated by the effect of the parasitic capacitances of the SMPS *e.g.* the heat sink-to-ground parasitic capacitance.

CONCLUSIONS

Based on a direct clamping two-probe measurement approach, the measurement procedure to extract the DM and CM impedance of SMPS and LISN consists in the following steps:

1. *Pre-measurement calibration process*: The setup impedances ($Z_{setup,DM}$ and $Z_{setup,CM}$) and the frequency-dependent coefficients (K_{DM} and K_{CM}) of the measurement setups must be determined according to the procedure outlined in Section **3**. Once these parameters are known, the measurement setup is ready to measure any unknown impedance using equation (10).

2. *Measurement of the noise termination impedances* ($Z_{LISN,DM}$ and $Z_{LISN,CM}$): As the SMPS is powered through the LISN, the LISN acts as a termination for the SMPS noise. To measure the LISN impedance, the SMPS will be replaced by a capacitor, which serves as an AC short circuit at high frequency. In the DM measurement setup, a 1 μF capacitor is connected between line and neutral as shown in Fig. **8** (**a**). In the CM measurement setup, a 1 μF capacitor is connected between the line and the ground (nodes L and G) and another 1 μF capacitor is connected between the neutral and the ground (nodes N and G) as shown in Fig. **8** (**b**). The DM and CM impedances of the LISN can be determined by means of equation (10).

3. *Measurement of the noise source impedances* ($Z_{SMPS,DM}$ and $Z_{SMPS,CM}$): In Figs. **12** (**a**) and (**b**), the measured impedance using the two-probe setup is the total impedance in the circuit loop; we designate as $Z_{T,DM}$ and $Z_{T,CM}$ such measured impedances. With the known setup impedance obtained from Step 1 and the known LISN impedance from Step 2, the respective DM and CM impedances of the noise source (SMPS) can be found according to (16) and (17).

As an example, the SMPS (VTM22WB, 15W, +12V$_{dc}$/0.75A, 12V$_{dc}$/0.5A) is powered through the LISN (Electro-Metrics MIL 5-25/2) and characterized by means of the setups shown in Figs. **12** (**a**) and (**b**). The DM and CM impedances of the LISN (noise termination) and the SMPS (noise source) are determined with Steps 1 to 3 described earlier. Figs. **16** (**a**) and (**b**) show the magnitudes and phases of the measured LISN and SMPS impedances in DM, respectively. Figs. **17** (**a**) and (**b**) show the magnitudes and phases of the measured LISN and SMPS impedances in CM. From Figs. **16** (**a**) and (**b**), the DM SMPS impedance magnitude is higher than the DM LISN impedance by a few tens ohms to a few hundred ohms and their phases are spanning approximately 90 degrees, over the frequency range of measurements. Figs. **17** (**a**) and (**b**) show that the CM SMPS impedance is capacitive and rather regular over the frequency range of measurements, while the CM LISN impedance shows a phase change not easily explainable in terms of elementary circuit elements.

With accurate magnitude and phase of termination impedances over the frequency range of interest, an appropriate EMI filter can be chosen, or it can be designed with optimal component values; hence, meeting specific conducted EMI limits becomes possible [27]. For example, based on termination impedance comparisons as shown in Figs. **16** and **17**, CL-filter configurations of both equivalent DM and CM filters might be chosen where the capacitors (X-caps and Y-caps) are at the SMPS side and the inductors (CM and DM chokes) are at the LISN side, as illustrated in Fig. **1** (**b**), because the capacitor, to be effective, must be placed in parallel to a high impedance and the inductor must be connected in series with a low impedance [28].

With the direct clamping two-probe measurement approach, the unknown impedances of the system under test can be extracted without any modification of the system. Unlike the impedance analyzer, this approach is not limited to measure the unknown impedances under power off condition. The measuring unknown impedances might be either powered or not. Moreover, the amplitude and phase of the DM and CM input impedances of a SMPS (noise source impedances) and the DM and CM output impedances of LISN (load impedances) under their actual operating conditions can be extracted with high accuracy. However, like any other measurement methods, the proposed method has some limitations which can be addressed as follows.

- For the results to be valid, it requires the condition where the injected signal is much higher than the background noise. For very high power active systems, additional power amplifier may be necessary to meet the mentioned condition.

- The accuracy of the proposed method can be ruined if the pre-measurement calibration process is not properly set. For example, the wire loop to the resistor-under-measurement must be made as small as possible to avoid any loop resonance below 30 MHz.

- To extract only the SMPS impedance, the power interconnection between LISN and SMPS must be as short as possible and is much shorter than the wavelength of the highest frequency of interest (30 MHz) to minimize any transmission line effects.

- The range of the unknown impedances under measurement must be roughly known so that the precision resistor R_{std} can be chosen properly. The R_{std} must be chosen somewhere in the middle of the range of unknown impedance to be measured.

REFERENCES

[1]   Clayton R Paul. Introduction to Electromagnetic Compatibility, John Wiley & Sons, second edition, 2006.
[2]   Reinaldo Perez, Handbook of Electromagnetic Compatibility, Academic Press, 1995.
[3]   Garry B, Nelson R. Effect of impedance and frequency variation on insertion loss for a typical power line filter, in 1998 Proc. IEEE EMC Symposium, pp. 691-695.
[4]   Specification for Radio Disturbance and Immunity Measuring Apparatus and Methods Part 1: Radio Disturbance and Immunity Measuring Apparatus, CISPR 16-1, 1999.
[5]   Tihanyi L. Electromagnetic Compatibility in Power Electronics. IEEE Press 1997.
[6]   Guerra A, Maddaleno F, Soldano M. Effects of diode recovery characteristics on electromagnetic noise in PFCs, in Proc 1998 IEEE Applied Power Electron. Conf., pp. 944-949.
[7]   Liu Q, Wang S, Wang F, Baisden C, Boroyevich D. EMI suppression in voltage source converters by utilizing DC-link decoupling capacitors. IEEE Trans Power Electron 2007; 22(4): 1417-1428.
[8]   Fluke JC. Controlling conducted emission by design, New York: Van Nostrand Reinhold, 1991.
[9]   Ferreira JA, Willcock PR, Holm SR. Sources, paths and traps of conducted EMI in switch mode circuits,in Proc. 1997 IEEE Industry Applications Conf., pp. 1584-1591.
[10]   Vakil SM. A technique for determination of filter insertion loss as a function of arbitrary generator and load impedances. IEEE Trans Electromagn Compat 1978; 20(2): 273-278.
[11]   Audone B, Bolla L. Insertion Loss of Mismatched EMI Suppressors. IEEE Trans Electromagn Compat Sep 1978; 20(3): 384-389.
[12]   Nave MJ. Power Line Filter Design for Switched-Mode Power Supplies, VNR, 1991.
[13]   Nagel A, De Doncker RW. Systematic design of EMI-filters for power converters, in Proc. 2000 IEEE Industry Applications Conf., pp. 2523-2525.
[14]   Caponet MC, Profumo F, Tenconi A, EMI filters design for power electronics. Proc. 2002 IEEE Power Electron. Spec. Conf., pp. 2027-2032.
[15]   Ye S, Eberle W, Liu YF. A novel EMI filter design method for switching power supplies. IEEE Trans Power Electron Nov 2004; 19(6): 1668-1678.
[16]   Schneider LM. Noise source equivalent circuit model for off-line converters and its use in input filter design, in 1983 Proc. IEEE EMC Symposium, pp. 167-175.
[17]   Zhang D, Chen DY, Nave MJ, Sable D. Measurement of noise source impedance of off-line converters. IEEE Trans. Power Electron Sep 2000; 15(5): 820-825.
[18]   Meng J, Ma W, Pan Q, Kang J, Zhang L, Zhao Z. Identification of essential coupling path models for conducted EMI prediction in switching power converters. IEEE Trans Power Electron Nov 2006; 21(6): 1795-1803.

[19]   See KY, Deng J. Measurement of noise source impedance of SMPS using a two probes approach. IEEE Trans Power Electron May 2004; 19(3): 862-868.

[20]   Tarateeraseth V, Bo Hu, See KY, Canavero F. Accurate extraction of noise source impedance of SMPS under operating condition. IEEE Trans Power Electron 2010; 25(1): 111-117.

[21]   Mitchell DM. DC-DC Switching Regulator Analysis, McGraw-Hil, 1988, ch. 4.

[22]   Brooks JL. A low frequency current probe system for making conducted noise power measurements. IEEE Trans Electromagn Compat 1965; 7(2): 207-217.

[23]   Southwick RA, Dolle WC. Line impedance measuring instrumentation utilizing current probe coupling. IEEE Trans Electromagn Compat Nov 1971; EMC-13(4): 31-36.

[24]   Nicholson JR, Malack JA. RF Impedance of power lines and line impedance stabilization networks in conducted interference measurement. IEEE Trans Electromagn Compat 1973; 15(2): 84-86.

[25]   Kwasniok PJ, Bui MD, Kozlowski AJ, Stanislaw SS. Technique for measurement of powerline impedances in the frequency range from 500 kHz to 500 MHz. IEEE Trans Electromagn Compat 1993; 35(1): 87-90.

[26]   David M Pozar. Microwave Engineering, JohnWiley & Sons, third edition, 2005.

[27]   Tarateeraseth V, See KY, Canavero F, Chang RWY. Systematic electromagnetic interference filter design based on information from in-circuit impedance measurement. IEEE Trans Electromagn Compat 2010; 52(3): 588-598.

[28]   Jasper J Goedbloed. Electromagnetic Compatibility, Prentice Hall, 1992.

CHAPTER 3

Heatsink EMI Effects in Power Electronic Systems

Gordana Klaric Felic[*]

National ICT Australia, Victoria Research Laboratory, The University of Melbourne, Parkville VIC 3010, Australia

Abstract: In power electronic systems, large transient phenomena excites the conducting structures and induces parasitic currents *via* heatsink stray capacitance thus causing radiated emissions. The key demands made by PCB mounted power circuits are related to low conduction losses, improved thermal performance, and lower inductance board layouts. Thermal management of power devices involves the use of heatsinks which are seen as parasitic elements from an EMI point of view. In order to reduce radiation the heatsink can be grounded at the cost of increased common mode currents to the power supply, escalating conducted EMI. Increased parasitic capacitance between the power device and the heatsink reduces the common-mode current but may upset the cooling efficiency of the heatsink. Thus, the connection of a heatsink to a power device and the packaging technology is a design issue involving EMI and thermal performance. The parasitic coupling path of a non-grounded heatsink can be mainly considered capacitive within the EMI regulated frequency range of 2 GHz. The common-mode coupling models for various heatsink configurations such as open frame, folded frame or PCB mounted configuration can be easily implemented in numerical computations by considering proper boundary conditions and energy sources. The heatsink models for numerical computations depend on the design, power level, switching frequency and packaging technology of the power electronic system.

Keywords: Analytical function of a complex variable, Antenna, Common mode coupling, DirectFET package, Finite Difference Time Domain (FDTD), Heatsink, Laplace's equation, Radiation pattern, Rectangular cavity, Switching transients.

INTRODUCTION

The increasing integration of power electronic circuit modules together with the continuing growth in power density, switching speed and operating frequency have resulted in a close interaction between electromagnetic, thermal and mechanical considerations and a significant increase in undesirable Electromagnetic Interference (EMI) effects. The high switching frequency operation of power electronic devices leads to increased heat generation due to switching losses since in real cases ideal switches do not exist. Fig. **1** illustrates the generation of switching losses during turn-on and turn-off switching time. During switching transients, there are significant switching losses associated with dv/dt and di/dt transients. These phenomena depend on several issues such as characteristics of power switches, control signals, gate drives, stray parameters and operating points of the system. Real power components have limited power, voltage and current-handling capabilities. They have also limited switching speed due to charging and discharging internal capacitors existing between the junctions which limit the maximum operating frequency of the device and create switching losses. *On*-state and *off*-state resistances or voltage drop and leakage current create conduction losses. When a switch is turned on or off, energy is lost during the switching transients when operating points of the switch are changed from on (off) to off (on) states through an active state.

This type of energy loss is called switching loss of the power switch. When a switch is off, normally a leakage current through the switch is very small and the energy loss associated with off-state can be ignored. But when the switch is on, the energy loss depends on the current through the switch and forward voltage of the switch. This type of energy loss is called conduction loss of the switch. The generated heat needs to be dissipated through a heatsink to keep the devices within safe operating conditions.

On the other hand, rapid switching capability of modern semiconductor devices (MOSFET, IGBT, MCT etc) reduces power loss. However, very fast voltage and current variations excite conducting metallic structures and induce parasitic currents *via* heatsink stray capacitance thus causing conducted and radiated EMI. Therefore there is a need for electromagnetic characterisation of conducting structures such as heatsinks and baseplates. In order to

---

*****Address correspondence to Gordana Klaric Felic:** NICTA, Electrical and Electronic Engineering, The University of Melbourne, Parkville VIC 3010, Australia; Tel: 613-83443804; Email: gordana.felic@nicta.com.au

Firuz Zare (Ed)

design a power electronic system with properly controlled electromagnetic interference, a designer typically uses a set of established EMC rules to estimate various electromagnetic parameters including heatsink EMI effects. The estimation of electromagnetic parameters is typically based on simplified analytical expressions which are often very crude approximations to the true physical situation [1]. This approach is valid for some applications but recent trends in power electronics toward integrated module and design optimisation necessitates the use of computer aided design tools for electromagnetic interference characterisation. In particular, EMI characterisation of heatsink effects in power electronic systems deal with coupling models that include modelling of physical structures and EMI sources.

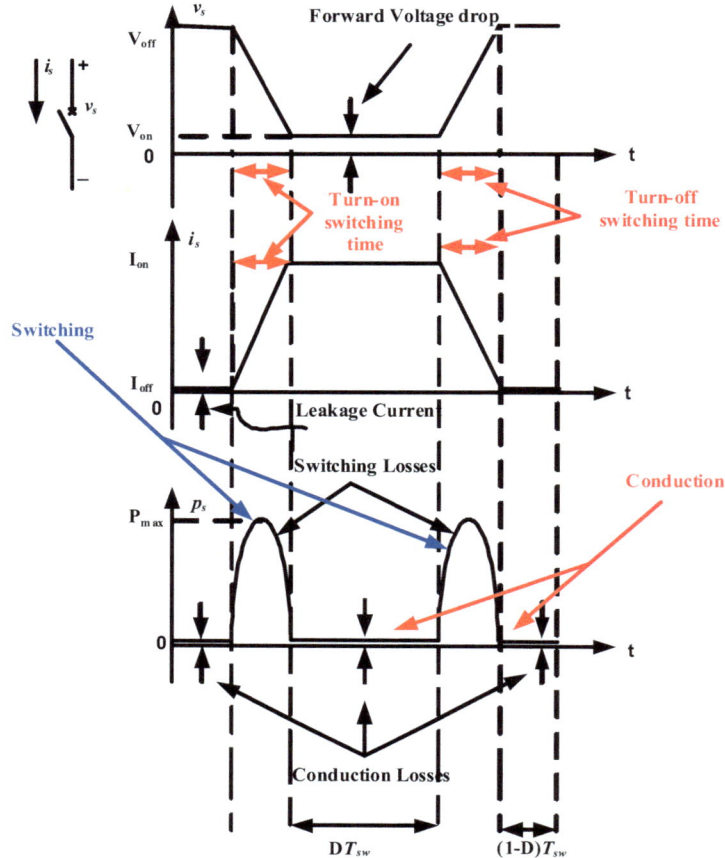

Figure 1: Generation of switching losses during turn-on and turn-off switching time.

Power heatsinks come in different geometries and sizes and operate with different power devices. Some typical heatsinks for power devices are shown in Fig. **2**.

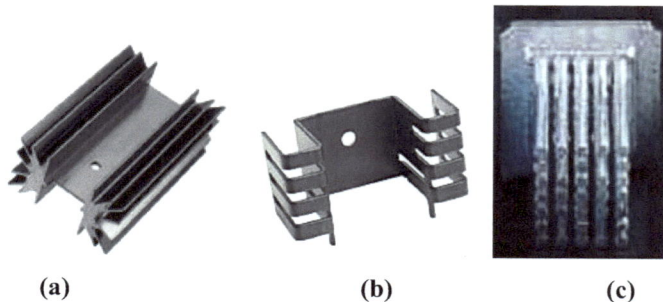

       (a)   **(b)**   **(c)**

Figure 2: Power Heatsinks: **a**) Extruded heat sink, Aavid Thermalloy **b**) Channel style heat sink with folded back, Aavid Thermalloy **c**) Miniature Heat Sink for DirectFET MOSFET, International Rectifier.

The EMI characterisation of power heatsinks leads to the numerical modelling of their radio frequency characteristics and a study of how the sizes, geometry and power package placement influences radiation patterns and resonance frequencies. Two typical power semiconductor packages TO-220 and DirectFET with a mounted heatsink are used for this analysis (Fig. **3**).

(a)   (b)

Figure 3: MOSFET power device packages: **a**) Infineon P/PG-TO200, **b**) International Rectifier - IRF6636TR1PBF - MOSFET, N, DirectFET.

TO-220 MOSFET is a style of electronic component package, commonly used for transistors, silicon-controlled rectifiers, and integrated circuits [2]. TO-220 packages commonly have three leads although units with two, four, five or seven leads are also manufactured. Components made in TO-220 packages can dissipate more heat than TO-92 packages which offer compact size at very small cost. The TO-220 packages are 'heatsinkable', and thus can be used for applications where large amounts of power are dissipated as heat. The top of the package has a metal tab with a hole used in mounting the component to a heatsink. Thermal compound is often applied to further improve heat transfer from the package to the heatsink. The metal tab is often connected electrically to the internal circuitry. This does not normally pose a problem when using isolated heatsinks, but an electrically-insulating pad or sheet may be required to electrically isolate the component from the heatsink if the heatsink is electrically conductive, grounded or otherwise non-isolated. Many materials may be used to electrically isolate the TO-220 package, some of which have the added benefit of high thermal conductivity. One should not touch a TO-220 package or heatsink when the specifics of the device are not known, to prevent risk of electrical shock or thermal burn. In applications that require a heatsink, damage or destruction of the TO-220 device due to overheating may occur if the heatsink is dislodged during operation. The TO-220 can be used in high-power and high-current applications where equivalent components of other cases may be susceptible to damage.

Recently there has been a move to reduce packaging related losses with introduction of a new technology, DirectFET that cuts conduction losses (reduces electrical resistance of the package) and improves cooling (reduces thermal impedance) [3]. The DirectFET package is fundamentally different to the TO-220 package since it dissipates heat from the die in opposite directions cooling through both PCB substrate pad and metal can on the top of the device. The International Rectifier DirectFET power package is a breakthrough surface-mount power MOSFET packaging technology designed for efficient topside cooling in a SO-8 footprint. In combination with improved bottom-side cooling, the new package can be cooled on both sides to cut part count by up to 60%, and board space by as much as 50% compared to devices in standard or enhanced SO-8 packages. This effectively doubles current density (A/in2) at a lower total system cost. The DirectFET MOSFET family offerings match 20V and 30V synchronous buck converter MOSFET chipsets, followed by the addition at 30V targeted for high frequency operation. The DirectFET MOSFET family is also available in three different can sizes giving maximum flexibility for different design needs. However, there are some disadvantages of DirecFET packages which are not attributes to TO-220 packages. In the case of chassis-mounting DirectFET, space must be dedicated to the components and/or their heatsinks, thereby increasing production costs. The DirectFET footprint requires considerably more surface area on a printed circuit board than other case styles, especially when heatsinks are used.

Fig. **4** Illustrates a heatsink mounted on the TO-220 package and a heatsink mounted on the DirectFET package.

TO-220

DirectFET

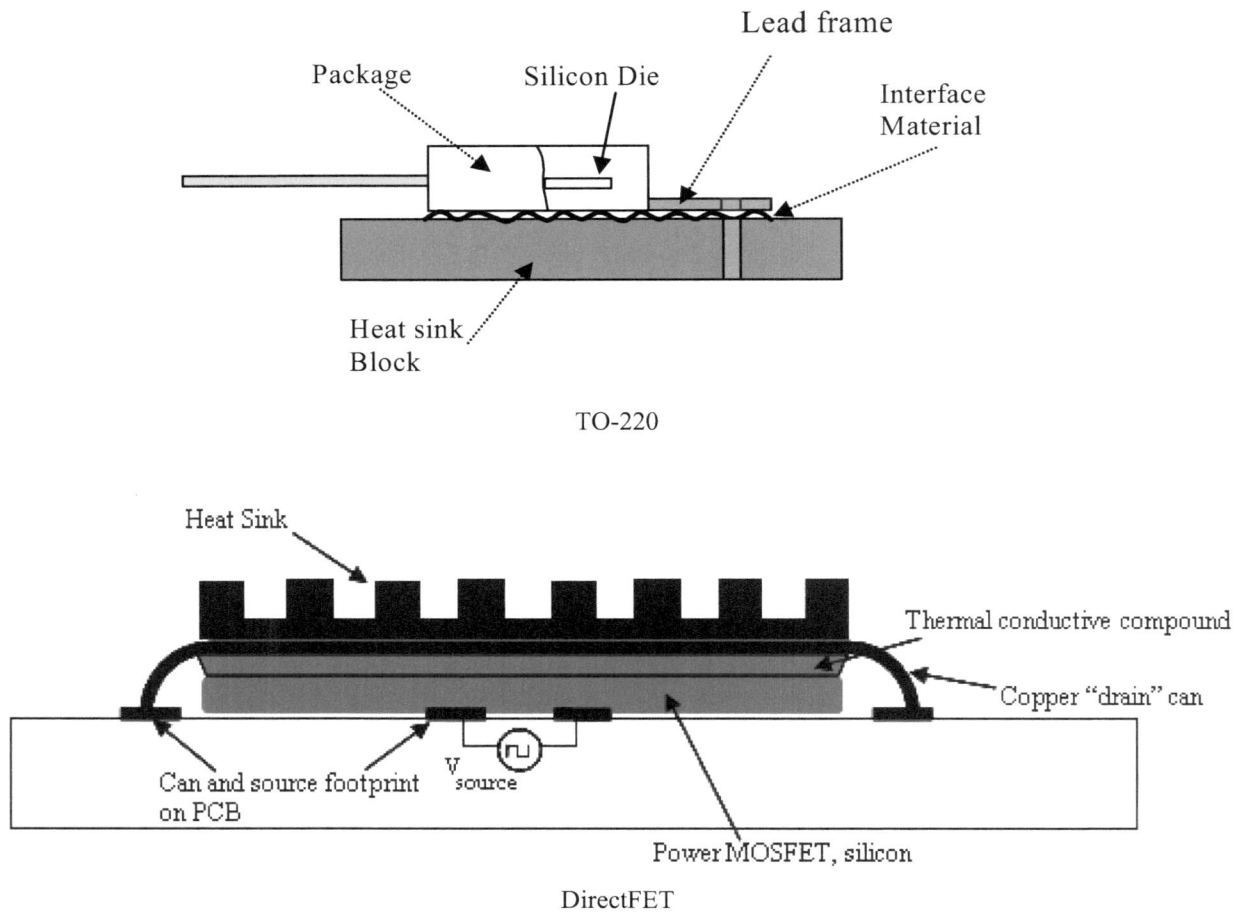

Figure 4: Power semiconductor packages with mounted heatsinks.

In the TO-220 package the heat sink is connected with lead frame by screws or clips and isolated from it by an air gap, a layer of thermal compound, or an isolator pad. The bottom of the chip (MOSFET drain) is connected directly to a lead frame and the top connection is made using bond wires. In the DirectFET package a heatsink is placed upon the metallic "drain' can of the DirectFET package. Since the can is soldered on the footprint of the printed circuit board and the total structure looks like a typical coupled microstrip line.

An example of typical heatsink arrangement in an AC/DC and DC-DC power conversions is shown below (Fig. **5**).

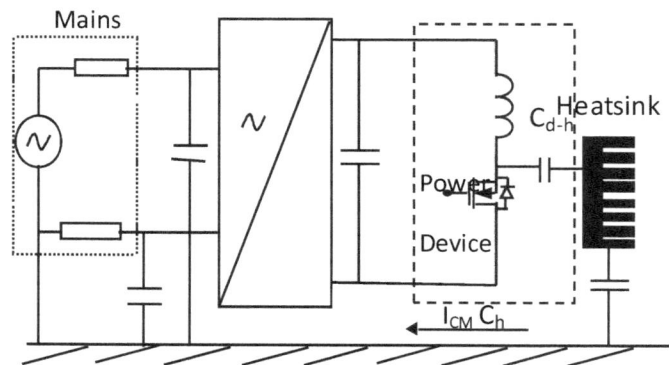

Figure 5: AC/DC converter with non-grounded heatsink.

The EMI issues related to the non grounded heatsink which forms capacitive coupling paths to ground has been discussed in [4]. In order to reduce radiation, the heatsink can be grounded at the cost of increased common-mode currents to the power supply, escalating conducted EMI. The non grounded and isolated heatsink reduces common-mode current but may affect the cooling efficiency of the heatsink. Thus, the connection of a heatsink to a power device and the packaging technology is a design issue involving EMI and thermal performance.

The modelling of the non-grounded heatsink in this chapter is based on coupling mode concepts suitable for numerical simulations *via* the Finite Difference Time Domain (FDTD) method [5]. The FDTD method can predict electromagnetic parameters over the resonance region in which a wavelength is comparable to the size of the object under study. It can work with a wide range of stimuli (disturbance sources), objects, environments and response locations. It is a 'volume' based time domain method that provides transient electromagnetic solutions for complex systems with a wide range of objects (conductors, dielectrics and metallic bodies). These capabilities make FDTD applicable to electromagnetic modelling of power electronic systems that operate with naturally generated transient effects. Power electronic systems are usually built with different objects such as conductors on PCBs and metallic structures such as heat sinks and base plates. Power semiconductor devices are disturbance sources represented as stimuli in FDTD that can be modelled as voltage or current sources. This is a key advantage of the FDTD method. In this chapter the modelling of disturbance source and heatsink structure using FDTD is explored in detail and some of the computational results are validated using analytical results and electric near field measurements. The main issues are the implementation of the heatsink problem in FDTD computational space and the selection of the energy sources. A solution to these issues leads to proper EMI characterization of heatsink EMI effects in power electronic systems.

HEATSINK PARASITIC MODES

The EMI caused by a heatsink is characterized as a common-mode EMI common-mode current which circulates through a coupling path between the power device, heatsink and ground. Fig. (**6**) gives an overview of the phenomena by considering the power device as an EMI signal that propagates *via* parasitic coupling paths (denoted by Z_{p1} and Z_{p2} impedances) to the heatsink and the reference ground. The Z_{p1} impedance depends on geometry of the connection between the power device and the heatsink and the Z_{p2} impedance depends on the coupling path between the heatsink and the ground. The parasitic coupling path of a non-grounded heatsink can be mainly considered capacitive within the EMI regulated frequency range of 2 GHz.

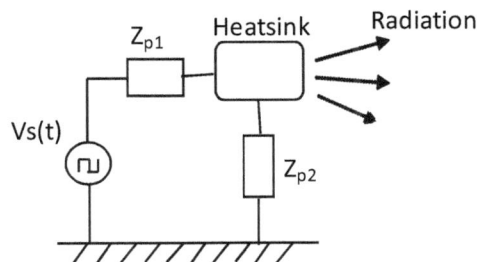

Figure 6: Schematic overview of the EMI effects associated with heatsink parasitic modes.

Here we create models that can be easily implemented in FDTD by considering proper boundary conditions and energy sources. The models depend on the design, power level, switching frequency and packaging technology of the power electronic system.

Open-frame configuration. A parasitic mode model relevant for power electronic systems with low power and low switching frequency and using standard packaging technology considers the non-grounded heatsink acting as a charged metallic structure or voltage-driven antenna (with respect to the reference ground). The model of a standard MOSFET package in an open frame configuration is illustrated (Fig. **7**).

The insulation between the metallic drain and heatsink forms the stray capacitance C_{d-h}. The common-mode coupling path to ground is formed by drain-ground capacitance C_d and heatsink-ground capacitance C_h. The energy source is a time varying drain-ground voltage. The heatsink and MOSFET drain are Perfect Electric Conductors (PEC). Therefore, the heatsink is seen as a voltage-driven antenna with respect to ground. Grounding the heatsink would decrease the radiating electric fields but increase the common-mode current (I_{CM}) back to the power mains.

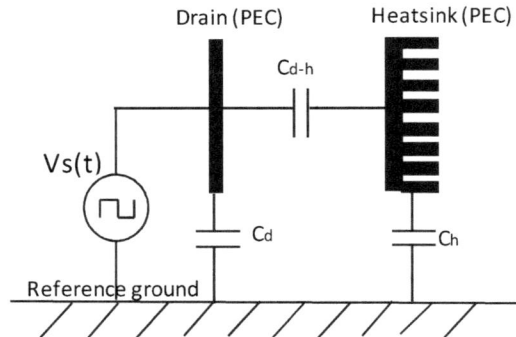

Figure 7: Common-mode Coupling Model (open-frame configuration)

Folded-frame Configuration. In low power supply applications (below about 100W and 100 kHz switch frequency) a folded-frame design (Fig. **8**) is used as a Faraday shield which prevents common-mode current ICM circulating from the heatsink into the equipment chassis.

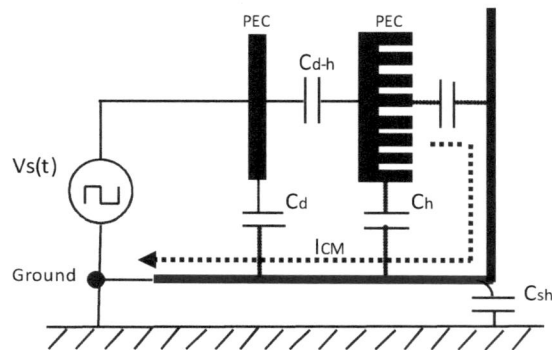

Figure 8: Common-mode Coupling Model (Folded-frame configuration).

In this configuration the common-mode coupling path to ground is formed by stray capacitances between the drain-heatsink structure and the equipment chassis. This configuration is better than the configuration with grounded heat sink since the common mode current is contained within the frame which is grounded and does not re-radiate. The stray capacitance between the shield and the reference ground adds to the total capacitance (C_h+C_{sh}) and reduces it due to series connection with C_h.

CLOSED FORM FORMULATION FOR STRAY CAPACITANCES

The total stray capacitance C_{total} as seen by source V_s is given by

$$C_{total} = \frac{C_{h-d}C_h}{C_{h-d} + C_h} + C_d \qquad (1)$$

where C_h is the stray capacitance of the heatsink, C_{h-d} is the drain-heatsink capacitance, and C_d is the drain capacitance. The voltage change (dV/dt) and the total stray capacitances of the system generate the common-mode current,

$$I_c(t) = C_{total} \frac{dV_s}{dt}$$ (2)

The stray capacitance between the power device and heatsink conducts the high frequency harmonics to the metallic surface of the heatsink. The distributed current harmonics on the heatsink surface generate near and far field EMI. Since the drain heat-sink capacitance C_{d-h} is much greater than the heat-sink to ground capacitance C_h, the total capacitance C_{total} mainly depends on the heat-sink ground and drain-ground capacitances ($C_{total} \sim C_h + C_d$). The stray capacitance can be estimated analytically by applying the theory of analytical functions of a complex variable to the metallic plate-ground electrostatic problem. To calculate the capacitance of the heatsink we use the known electrostatic potential formulation in angular space described in [6] and then extent it to the heatsink 3-D structure problem (Fig. **9**).

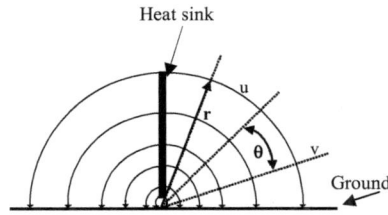

Figure 9: Two plate PEC structure model.

The potential between the conducting plates (heatsink and ground plane) (Fig.9), which are kept at potential V_1 and V_2 and make an angle $\pi/2$ can be found by solving Laplace's equation [7]. From the shape of the plates it follows that the electrostatic potential depends on angle θ. The electrical force between the charges plates is governed by Coulombs law. This force is the gradient of electrostatic potential and at any point between the plates (free of charges) $v(x,y)$ is a solution of Laplace's equation $\Delta^2 v = 0$. The theory of analytical functions of a complex variable $f(z) = f(x + jy) = w = u(x, y) + jv(x,y)$ can be used to describe mathematical properties of static electric fields where w is the value of f at $z = x + jy$. Thus, function f is called the complex potential corresponding to the real potential v and u corresponding to lines of electric force (flux lines):

$$f(z) = w = F_1 \ln \frac{z}{c} + F_2 + jF_3$$ (3)

where F_1, F_2 and F_3 are constants. The electric flux (u) is a function of r and potential is a function of angle θ

$$u = F_1 \ln \frac{r}{c} + F_2$$ (4)

$$v = F_1 \theta + F_2$$ (5)

If u represents an electric flux function then electric charge density along r is given by

$$\lambda = \varepsilon(u - u_1) = \varepsilon(F_1 \ln \frac{r}{c} - F_1 \ln \frac{r_1}{c})$$ (6)

If v represents equipotent lines then for $\theta = 0$, $v = V2$ and for $\theta = \frac{\pi}{2}$, $v = V1$ where V_1 is the electric potential of the heatsink plate and $V2$ is electric potential of the reference ground plane:

$$V_1 = F_1 \frac{\pi}{2}$$ (7)

$$F_1 = \frac{2V_1}{\pi}$$

(8)

Then the electric potential function is given by:

$$v = \frac{2V_1}{\pi}\theta$$

(9)

and the electric flux function is given by

$$u = \frac{2V_1}{\pi}\ln\frac{r}{c}$$

(10)

The surface charge density σ can be derived from the line charge density defined by the flux function as:

$$\lambda = \varepsilon(u - u_1) = \frac{2V_1\varepsilon}{\pi}\ln\frac{r}{r_1}$$

(11)

and

$$\sigma(r) = \frac{d}{dr}\lambda(r)$$

(12)

$$\sigma(r) = \frac{d}{dr}(\varepsilon\frac{2V_1}{\pi}\ln\frac{r}{r_1}) = \frac{2}{\pi r}\varepsilon V_1$$

(13)

Then the total charge Q can be calculated from the surface charge density of the surface S that represents the side walls of the plate heatsink as

$$Q = 2\oint_S \sigma dS$$

(14)

and the stray capacitance of the side walls

C_{h-w} can be calculated from

$$C_{h-w} = \frac{1}{V}\oint_S \sigma dwdr + \frac{1}{V}\oint_{S_1} \sigma_1 dS_1 = 4\int_0^w dw\int_{r_1}^r \varepsilon\frac{1}{\pi r}dr$$

(15)

The bottom side capacitance of the heatsink C_{h-b} can be calculated from the parallel plate capacitance formula:

$$C_B = \varepsilon_0\frac{S_1}{r_1}$$

(16)

where S_1 is the surface and r_1 is the distance between the plate and the ground plane. The contributing capacitance of the transition region C_{h-F} (fringing field at the edges of the plate) between the uniform field (bottom) and the logarithmic field of the sides is derived from a formulae that is based on the Schwartz-Christoffel's transformations, [8]:

$$C_{h-F} = 0.88\varepsilon_0$$

(17)

so that the expression for the total stray capacitance of the heatsink becomes

$$C_{h-Total} = 4\int_0^w dw\int_{r_1}^r \varepsilon\frac{1}{\pi r}dr + \varepsilon_0(\frac{S_1}{r_1} + 0.88)$$

(18)

For example if we consider a simple heatsink (Fig. **9**) where the width of the plate $w = 20$ mm, the distance between the heatsink and the conducting plane $r_1 = 4$ mm and the height of the heatsink $r = 15$ mm, then the estimated stray capacitance of the heatsink in air is $C_{Total} = 0.6$ pF.

NUMERICAL SIMULATIONS TO ANALYZE EMI EFFECTS

The FDTD method was first introduced in [9] and later developed in [10]. The rapid advances in computer technology have made the FDTD method increasingly convenient for use in electromagnetic computations. The method is based on the mathematics of finite-difference approximations to Maxwell's equations for electromagnetic field solutions in the time domain. The FDTD technique solves for electric and magnetic fields that are discreticised over rectangular grids together using a finite difference approximation of the spatial and temporal derivatives appearing in the differential form of Maxwell's equations.

FDTD can predict electromagnetic parameters over the resonance region in which a wavelength is comparable to the size of the object under study. It can work with a wide range of stimuli (disturbance sources), objects, environments and response locations. It is a volume based time domain method that provides transient electromagnetic solutions for complex systems with a wide range of objects (conductors, dielectrics and metallic bodies). These capabilities make FDTD applicable to electromagnetic modelling of power electronic systems that operate with naturally generated transient effects. Power electronic systems are usually built with different objects such as conductors on PCBs and metallic structures such as heat sinks and base-plates. Power semiconductor devices are disturbance sources represented as stimuli in FDTD that can be modelled as voltage or current sources. This is a key advantage of the FDTD method.

Thus, the way to better account for the heatsink structure is by using a three dimensional numerical computations where all the electric and magnetic fields can be calculated through a volume containing the heatsink. This section demonstrates how numerical simulations are used for modelling and analysis of heatsink electromagnetic behaviour for TO-220 and DirectFET package configurations shown in the introduction (Fig. **10**).

Computational Model. A heatsink can be modelled as a conductive rectangular volume (conductive solid box) or as a conducting patch (microstrip model). The solid box is typically positioned on the top of the chip package mounted on a PCB with a ground plane. A simple geometry of a heatsink is a parallel-plate structure representing the drain and heatsink (Fig. **10**). It is placed inside the FDTD computational domain with total dimensions IΔx x JΔy x KΔz (where I, J, K are integers that define the FDTD space size and Δx, Δy and Δz define the cell size). In order to ensure computational stability the maximum allowable time step size Δt must be limited so that a wave does not propagate through more than one cell at a time. This condition known as the Courant - Friedrichs - Lewy condition [11] determines that for computational stability:

$$\Delta t \leq \frac{1}{v_{max} \sqrt{\dfrac{1}{\Delta x_{min}^2 + \Delta y_{min}^2 + \Delta z_{min}^2}}} \tag{19}$$

where v_{max} is the maximum propagation speed of the wave inside the medium and Δx_{min}, Δy_{min} and Δz_{min} are the minimum cell sizes in each of the three directions.

The dielectric material between the plates forms a thin interface. The heatsink plates are set as aluminium electric conductor material. The ground plane, the lowest plane of the mesh edges is simulated as a perfect electric conductor (PEC boundary conditions). The other five walls of the computational space (cube) are simulated as an unbounded space with Liao absorbing outer boundary conditions [12].

An alternative is to use an Absorbing Boundary Condition (ABC) on all six walls assuming that the entire system floats in open bounded computational space, (Fig. **11**). All walls of the computational space (cube) are simulated as an unbounded space applying absorbing outer boundary conditions.

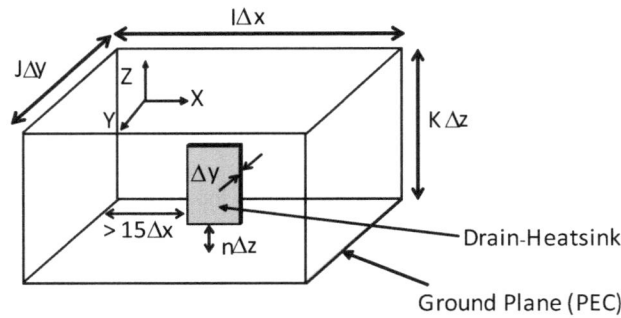

Figure 10: Heatsink in computational space with PEC and Liao boundary conditions.

The ground is located $n\Delta z$ cells from the parallel-plate structure and set to a perfect electric conductor. In order to minimise the reflections due to truncation of the computational space the distance from the heatsink to the walls is set to more than 15 cells ($>15\Delta x$, $>15\Delta y$ and $>15\ \Delta z$), [11]. The condition given by Eq. 19 ensures that the reflections from the walls after propagation through 15 cells are minimal.

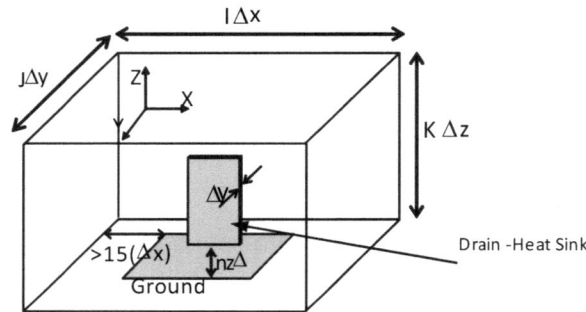

Figure 11: Heatsink in FDTD geometry with ground plane and open boundary conditions.

Energy Source. An accurate prediction of the radiation level depends on a proper physical correspondence with the excitation source. One of the most often used models in FDTD is an added voltage source between the ground plane and the heatsink. This is implemented by adding a predetermined source function to the naturally evolving electric field at a precise location [11]. The source selection depends on the previously determined coupling models of a heatsink and the required complexity of the EMI investigation. There are two approaches that may be used for the representation of internal energy sources in FDTD: a) the drain voltage referenced to the ground plane, b) the drain-source voltage of the power device. The internal drain voltage is represented by an added voltage source applied across a single rectangular cell oriented in the z direction and located between the ground plane and drain structure. The source location is selected by specifying the coordinates ($i_s\Delta x$, $j_s\Delta y$, $k_s\Delta z$) within the FDTD space. The voltage source has an internal resistance integrated into the FDTD code (Fig. 12). The resistance serves two purposes; it acts as a series internal resistance when a voltage source is activated and it acts as a termination when the source is deactivated.

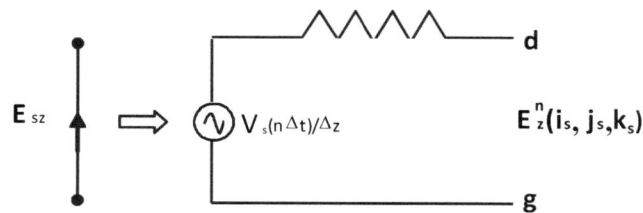

Figure 12: A z-directed voltage source with internal resistance.

$V_{d-g}(t)$ is changed when the voltage source V_s and the electric field change. At every time step, the field values are automatically updated with the $(n+1/2)^{th}$ current according to the explicit FDTD scheme that includes the voltage source [11].

$$(I_s^{n+\frac{1}{2}})_{i_s,j_s,k_s} = \frac{V_{d-g}^{n+1} + V_{d-g}^n}{2R_s} + \frac{V_s^n}{R_s} \tag{20}$$

A model of a practical heat sink mounted on a power device involves the geometry of the device package (TO-220 or DirectFET) with drain-source or drain-ground voltage excitation. The drain-ground voltage of the power device is represented by an added voltage source (V_s), applied across a single rectangular cell oriented in the z direction and located between the drain and the ground (Fig. **11**). The orientation of the voltage source depends on the power device package and the PCB layout.

Boundary Conditions. The boundary conditions influence electric field distributions of the heatsink. The application of PEC boundary condition reduces computational time but results in a less realistic field distribution. A simple drain heat-sink model with ground plane in FDTD computational space is shown below (Fig. **13**).

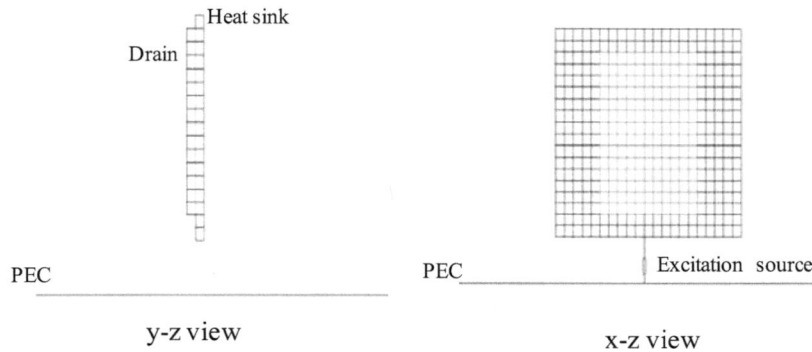

Figure 13: Heatsink and drain structure with PEC boundary condition at z=1.

Example. The FDTD computational space is a rectangular mesh represented by 80, 40 and 40 cells in x, y and z direction respectively. The size of a single cell in x, y and z direction is 1mm x 1mm x 1mm. The drain is represented by a rectangular PEC of 14 x 11 x 1 cells and the heatsink is represented by a PEC rectangular structure 17 x 11 x 1 cells. The reference ground is modelled by PEC boundary conditions or by a plane PEC structure. An infinitely thin dielectric sheet ($\varepsilon_r=4$) is also placed between the drain and heatsink. The above simple geometry is applicable to TO-220 package and it is used to demonstrate the effect of boundary conditions on modelling results.

The electric field in an x-z plane (y=50 cells) located at t=1ns when the excitation voltage source reaches its peak is shown in (Fig. **14**). The electric field flux distribution is in accordance with the electrostatic model of the conducting metal plate located above the ground plane.

If a PCB is included in the simulation then open boundary conditions are more appropriate to use since the entire PCB structure then becomes radiating object. Fig.**15** shows a thin PEC ground plane on 2 mm thick FR4 board 50 x 50 cells in size. In the vicinity of the heatsink the electric field intensity decreases significantly when the distance increases (Figs. **14** and **15**). The field intensity at 2 cm to the drain-heatsink structure is 200 V/m and it increases to 800 V/m at 1 cm distance from the heatsink.

As it is expected the size of the heatsink (width and length) affects the field intensity in the near field region.

The heatsink size increase in width from 11 to 22 cells and length from 17 to 34 cells increases the electric field intensity (Fig. **16**). The field intensity is at 2 cm to the drain-heatsink structure is above 300 V/m and it increases to 1.2k V/m at 1 cm distance from the heatsink.

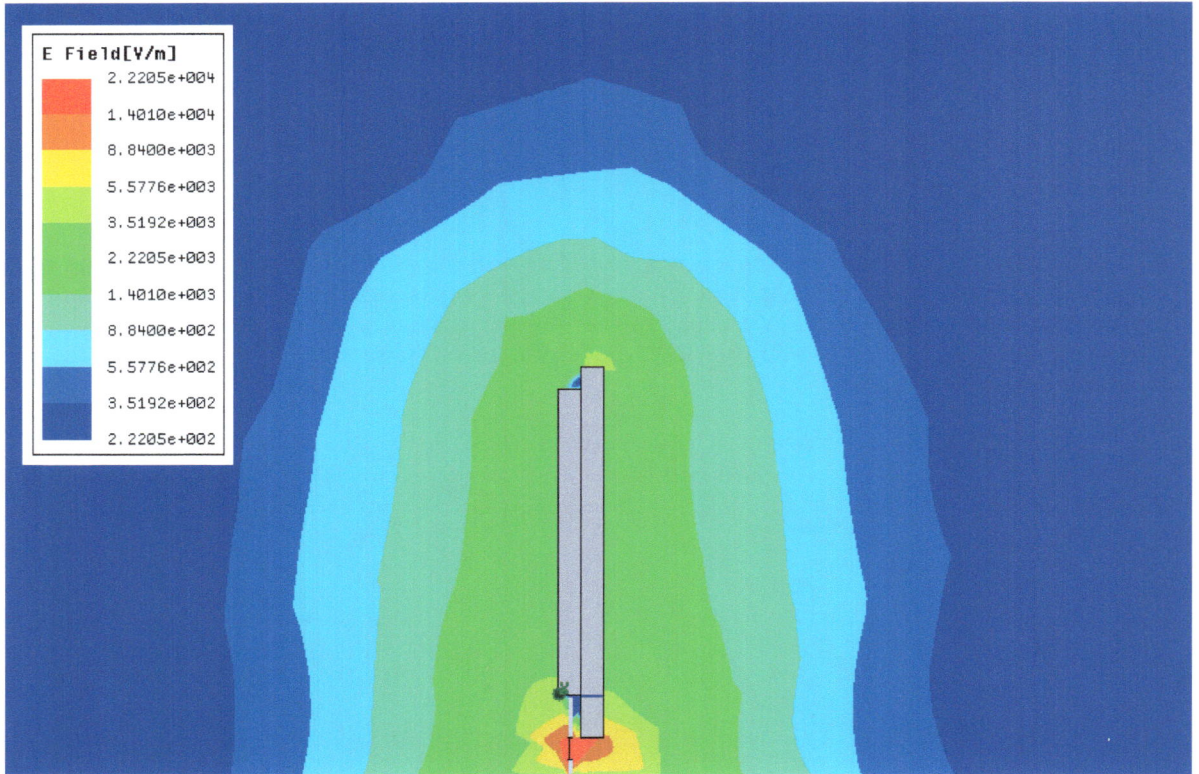

Figure 14: Electric field map of drain-heatsink with PEC boundary conditions.

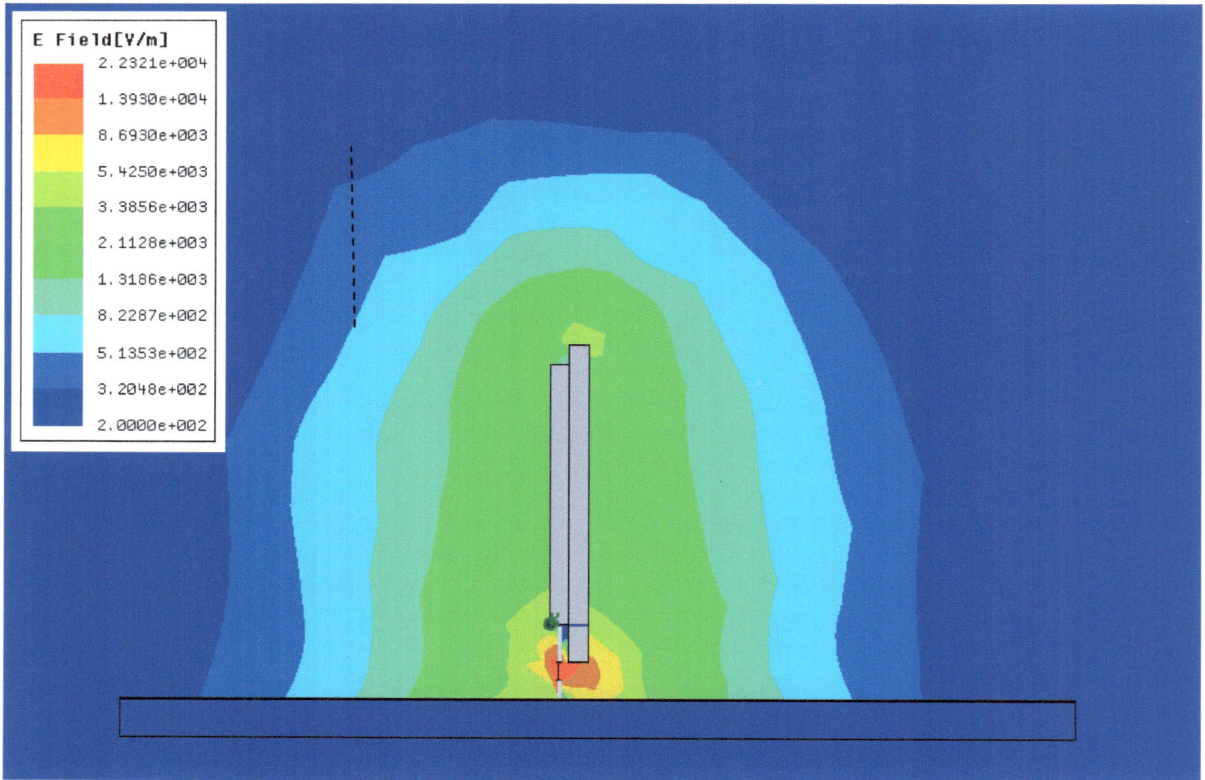

Figure 15: Electric field map of drain-heatsink on PCB with open boundary conditions.

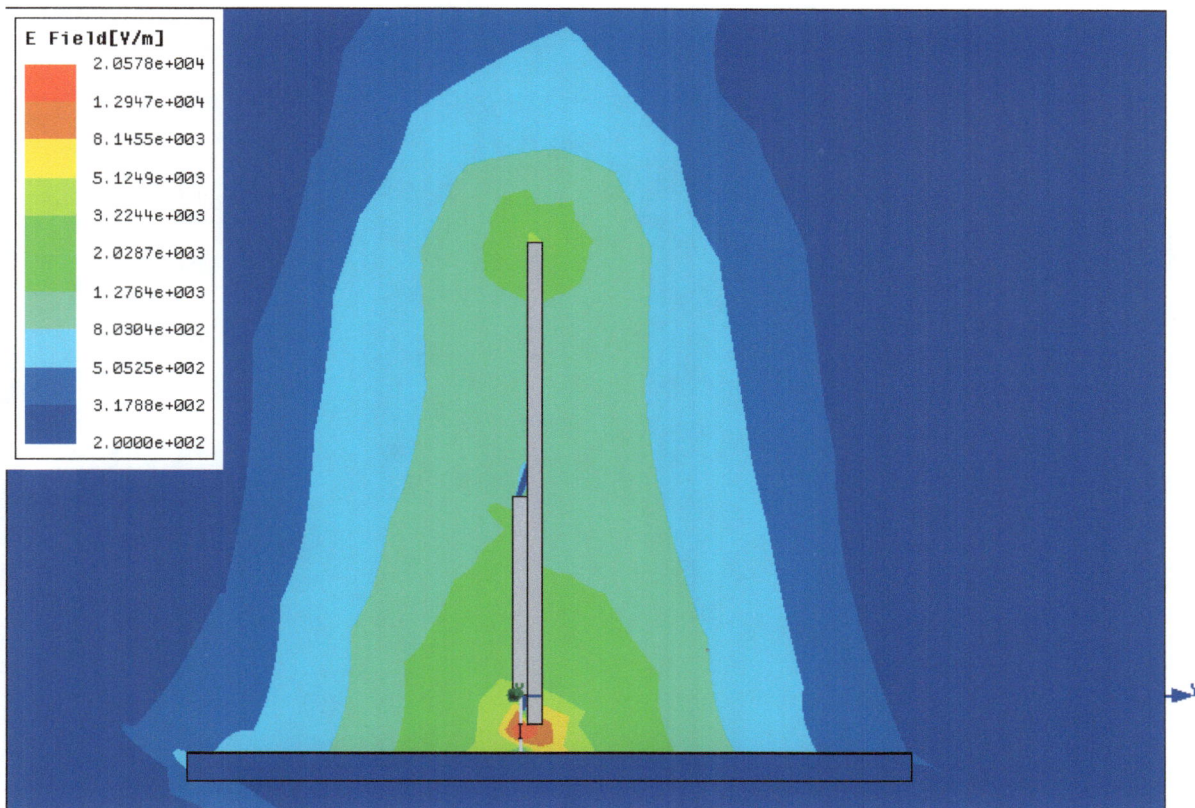

Figure 16: Electric field map of the large heatsink.

The second example is the heatsink mounted on the DirectFET package. As it is mentioned in the introduction a heatsink is placed upon the metallic "drain' can of the DirectFET package. Since the can is soldered on the footprint of the printed circuit board and the total structure looks like a typical coupled microstrip line. Like in the previous example with TO-220 package the same heatsink structure and open boundary condition are used. The metal can at the top of the device and the fitted heatsink act as an electric shield sustaining the electric field inside the package and radiating through the PCB substrate (Fig. **17**). The electric field above the heatsink is very low, below 12 V/m.

From the Electromagnetic Compatibility (EMC) aspect the DirectFET package obviously provides a better design solution than the TO-220 package. The electric field in the vicinity of the package is low and most of the field is contained at very close distance to the package. This potentially reduces the near field coupling to the circuit or trace placed in the vicinity of the power device (DirectFET package). However, it takes more space on the board due to the horizontal mounting. If the board space is of concern for the designer then the use of DirectFET package can be a significant limitation. The use of the folded shield with TO-220 package may provide a solution which limits the near field radiation and also does not significantly enlarge the size of the entire system.

ANTENNA

The primary criterion in selecting heatsinks for power electronic circuits is their thermal characteristics. However, the use of heatsinks can increase far-field EM radiation that generates emissions up to 1-GHz. Therefore given a choice of two or more heatsinks having similar thermal characteristics, the EMI considerations addressed in the design stage may be useful in mitigating radiated emissions. In this section the electromagnetic modelling of the heat sink is extended to "antenna like" models for prediction of far-field EMI. The far-field region is defined as "that region of the field of an antenna where the angular field distribution is essentially independent of the distance from the antenna. If the antenna has maximum overall dimension D, the far-field region is commonly taken to exist at distance greater than $2D^2/\lambda$, λ being the wavelength", [13]. This criterion is used for "surface-type antennas" such as patch antenna and heatsink structures. For example if the overall size of the heatsink is 5 cm then the far field region distance at 1 GHz frequency is 1.66 m.

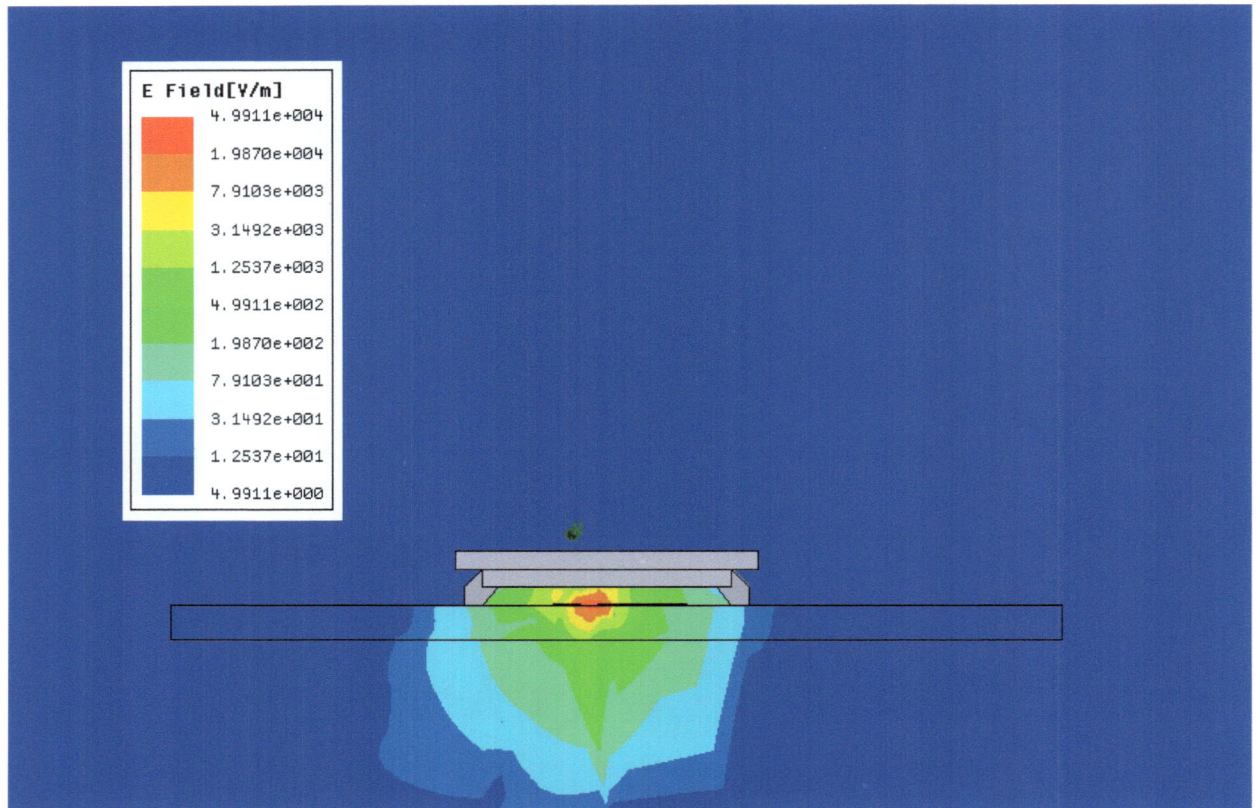

Figure 17: Electric field map of DirectFET package with heatsink (17 x 11 cells) on PCB.

Radiation pattern computations are performed at frequencies specified by the excitation source and represent the electric field values computed for a number of angles θ, between the polar axis z and a vector in the direction to the observation point (Fig. **18**). The distance from the centre of the heatsink (source cell) to the far-zone field observation point is 3 m (a distance defined by EMI emission standards), so that in order to calculate fields at this distance, the Huygen principle [14] is used. The fields are calculated inside the computational region, reproduced on the Huygen's surface located inside the region and then the radiation patterns of a Huygen's surface are computed and plotted in the spherical coordinate as shown (Fig. **18**).

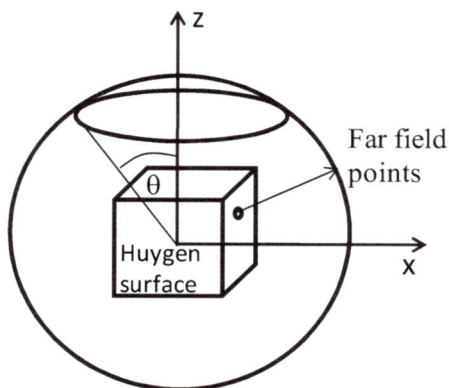

Figure 18: Coordinate system for antenna analysis.

The radiation (E-field intensity) patterns are generated from the common mode coupling model shown in Fig. **19** using APLAC simulation software [15]. The heatsink is placed on the component package where the parasitic capacitance between the transistor drain and the heatsink exists in the form of a dielectric insulator. The added excitation source represents the drain-source voltage harmonic in x or z direction that is applied across a single rectangular cell. The heatsink is in free space and floating with no connection to the reference plane and the components like leads and pins are neglected. For this arrangement the x-directed excitation source is more realistic representation of the actual EMI source (drain-source voltage) than z-directed excitation source which is directly connected from the ground plane to the heat sink (Fig. **13**).

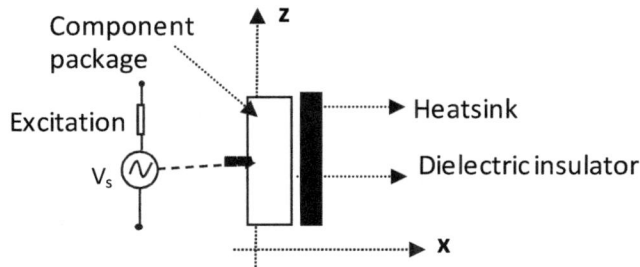

Figure 19: Antenna models (common-mode coupling) for numerical simulations.

Fig. **20** shows the radiation patterns of the heatsink at 200 MHz and 1 GHz with the x-directed excitation source Vs. The far field distance from the heatsink to the observation point is 3m. The E-field pattern values are in μV/m for 200 MHz radiation pattern (r=1μV/m) and mV/m for 1 GHz radiation pattern (r=3mV/m, 2mV/m, 1 mV/m). A heatsink with source excittaion of 1 V (Fig. **19**) at 200 MHz radiates 40 dB(μV/m) signal. The CISPR radiated interference standard electric field strength limit is 36.5 dB(μV/m) at 200 MHz and 37 dB(μV/m) at 1 GHz.

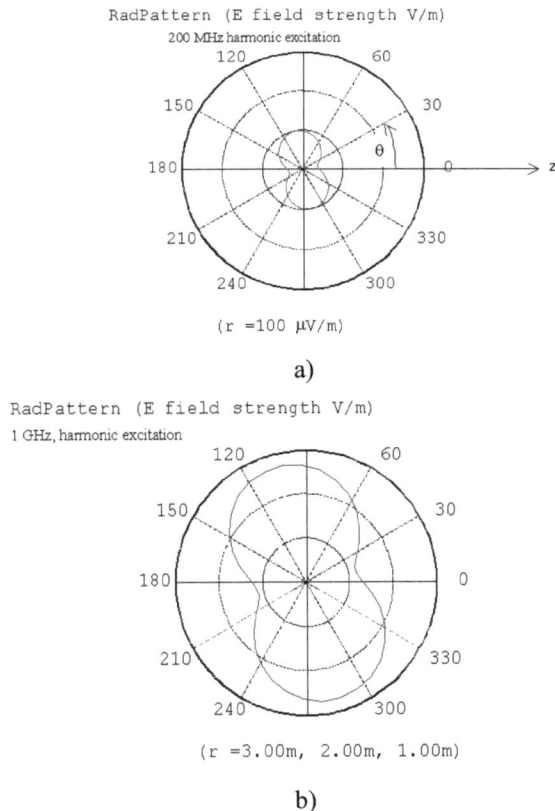

a)

b)

Figure 20: Radiation patterns of the heatsink: a) 200 MHz; b) 1 GHz, [16].

A summary of the radiation pattern results in terms of maximum electric field intensity are obtained for the heatsink modelled as a rectangular metallic slab 12mm long, 8mm wide and 6mm high placed on the top of the component package and for a larger heatsink, 30mm long, 25mm wide and 20mm high. The same computation was also carried out for a smaller heatsink with fins and for a switching transistor without a heatsink.

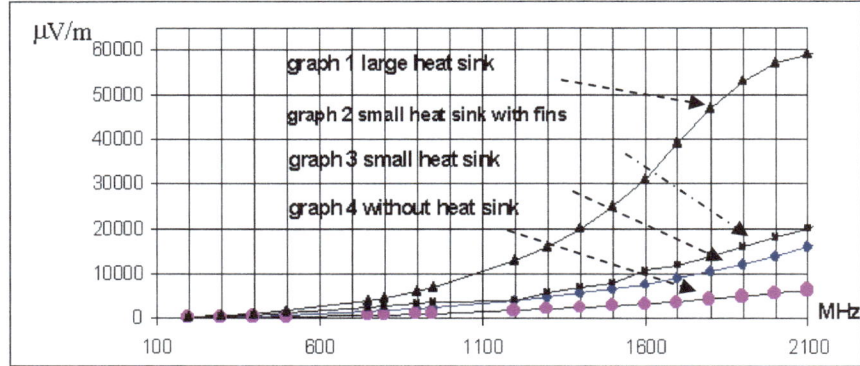

Figure 21: Maximum radiation intensity versus frequency with excitation source amplitude of 1V, [16].

Generally larger sized heatsinks produce more intensive electromagnetic fields. The field intensity is not linearly dependent on the frequency and heatsink dimensions, as shown (Fig. **21**). Heatsinks of particular dimensions at particular frequencies become very efficient radiators. This behavior can be explained by the resonance frequencies in an empty rectangular cavity which can be calculated by the following formula [13]:

$$(f_r)_{mnp} = \frac{c}{2\pi}\sqrt{\left(\frac{m\pi}{W}\right)^2 + \left(\frac{n\pi}{L}\right)^2 + \left(\frac{p\pi}{h}\right)^2} \tag{21}$$

where c is the speed of light and m, n, p represent the number of half-cycle field variations along the x, y, z directions, respectively. The same formula can be applied for the calculation of resonance frequencies of rectangular heatsinks where W is the length, L is the length and h is the height of the heatsink.

The radiated fields of the large heatsink increase for frequencies up to 2 GHZ (Fig. **21**). The field wavelength at 2 GHz is 150 mm which is much greater than the heatsink dimensions in this example and therefore the resonance does not occur at these frequencies. The field increase below 500 MHz is slow compared to the field increase at frequencies above 500 MHz. The effect of fins mounted on the small heatsink is not significant in this frequency range due to its smaller dimensions, much smaller than the wavelength at these frequencies. However, the addition of fins to the heatsink can reduce the resonant frequency and slightly increase value of the electric far-field radiation if they run parallel to the larger side (width or length). High power heatsink dimensions can be much larger and thus might radiate significantly even at frequencies less than 1 GHz. Due to the increasing switching speed of SMPS, and consequently the increasing higher order harmonics, the use of heatsinks may result in EMC compliance problems. The far field radiation at very high frequencies, above 200 MHz and up to 2 GHz may exceed EMC standard levels.

The above given radiation patterns show that the heatsink radiates as a directional antenna with maximum at θ =90° to the elevation plane. The radiation can be minimised by paying attention to the position of the heatsink with respect to the power device and the radiation patterns can be used to compare the effects of different heatsink positions. The excitation source which represents a power device can be mounted at specific points on the heatsink.

EMI DIAGNOSTIC MEASUREMENTS

The most effective EMI reduction technique in power electronic design is to directly control the main EMI sources. Two main design targets low power loss and low radiated magnetic fields in the vicinity of the converter become achievable by controlling the rate of variation of the drain current in the converter. The design method introduced in

[17] and [18] controls the slopes of the drain voltage and current by shaping the gate current in the main power switching devices. Since the power losses and switching transients largely depend on the slope of the drain current, this method results in faster drain voltage slopes without a consequent change in the current slope. The faster drain voltage slopes reduces the switching power loss while the amplitudes of radiated magnetic fields remain unchanged, which depend on the drain current. This method achieves increased efficiency without any change in magnetic field radiation. However, all types of radiated EMI covering magnetic fields, electric fields and far field radiation can be expected from power electronic circuits covering a wide frequency range from DC-1GHz. Therefore other aspects of EMI which are related to the effects of drain-source voltage variation and parasitic elements such as heatsinks should be considered in the power electronic design.

The starting assumption is that the drain source voltage variations are the main source of common mode based EMI and use of a heatsink in a SMPS circuit increases the electric field intensity and radiation in the vicinity of the circuit. Therefore a measurement system is developed to determine the intensity of the electric fields and their frequency spectra with and without a heatsink. Generally, the purpose of the diagnostic EMI measurements is to answer the following questions:

a)   Does the heatsink increase the amplitudes of the radiated electric fields in the vicinity of a SMPS circuit?

b)   How does dV_{d-s}/dt affect the amplitude of the radiated electric field measured in the frequency domain?

c)   In what frequency range does the amplitude of the electric field radiated by a SMPS become significant?

Experimental Circuit. In order to answer the above questions and demonstrate the validity of the concepts relating to heatsink radiation, a simple DC-DC buck converter was used. The converter shown in Fig. **22** consists of a gate controlled MOSFET, IC drive circuit, freewheeling diode, inductor and filter capacitors all located on a PCB. The heatsink mounted on the power MOSFET has overall dimensions of 30mm, 25mm and 20mm (width, length and height).

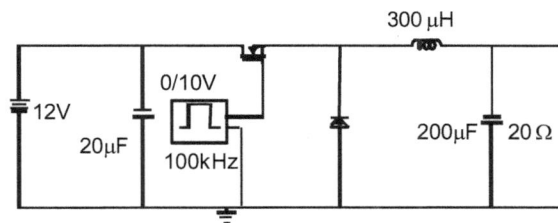

Figure 22: DC-DC buck converter.

Results. Measurements of near and far field signals were taken by using a diagnostic E-filed monopole probe (rod antenna) in conjunction with a spectrum analyzer. This probe is used to locate and quantify EMI sources and cover a frequency range from 2 MHz to 2000 MHz. Typical sensitivity of the probe is -10 dBm(V/m). The experimental results in the figures below show the electric filed emanating from the DC-DC buck converter circuit measured by an electric field probe connected to a spectrum analyser. The power of the signal is measured in the frequency domain so that the amplitudes are presented in [dBm]. The measurements were taken at distances of 2cm (near field for the frequencies from 20 MHz to 200 MHz) and 1m from the MOSFET for frequencies ranging from 200 MHz to 1000 MHz. A particular attention was paid to placement of the probe ensuring the same position during each measurement. The power level of the DC-DC buck converter used in this experiment was below 10 W with a switching frequency above 100 kHz. A common expectation is that the radiation at these frequencies becomes a greater concern forcing designers to consider shielding techniques in order to reduce EMI [1].

From the first set of measurements taken for the SMPS without a heatsink mounted on the power switching device it can be observed that the electric field signal power spectra are mainly in the range from 20-120MHz as shown in Fig. **23**. By comparing the signals from Fig. **23** with the signals measured when the converter is off, it is evident that the DC-DC buck converter radiates field signals 10-30dB above the existing background noise level (-50 dBm). We have accounted for the fact that some of the radiated E-field signals are caused by the drive circuit and power supply circuit.

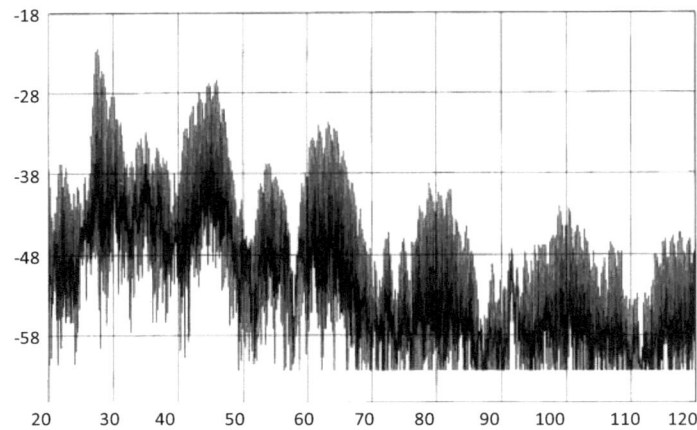

Figure 23: E-field signals of the SMPS without a heatsink, [dBmW / MHz], [19].

The next step in the experiment was to mount the heatsink on the top of the transistor and measure the E-field signal keeping the E-field probe in the same position, (Fig. **24**). When the heatsink is connected and isolated from the power-switching transistor, the E-field signal frequency content changes significantly.

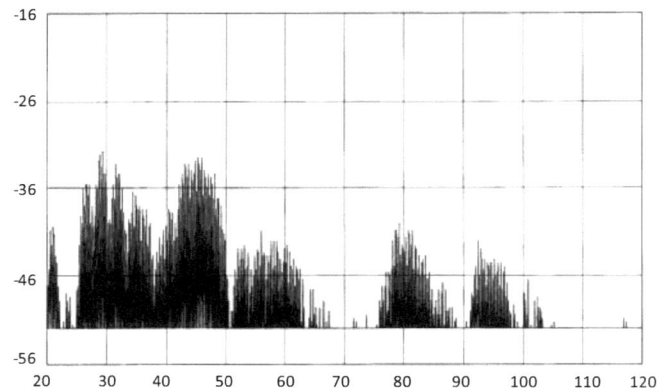

Figure 24:. E-field signals of the SMPS with a heatsink, [dBmW / MHz], [19].

By comparing Figs. **23** and **24** we see that the heatsink has a considerable effect on the frequency content of the E-field signal. Firstly, the heatsink causes broader frequency spectra in the ranges 30-40MHz, 40-50 MHz and 50-60 MHz. Secondly, the spectrum amplitudes in the frequency range below 30 MHz decrease or remain unchanged when the heatsink is present. This can be due to the relatively low stray capacitance (C_{d-h}) which does not conduct the frequency harmonics lower than 30 MHz. Thirdly, power spectra amplitudes of E field signals in the frequency range above 60MHz generally decrease when the heatsink is present.

The same measurements are presented for the same SMPS but with increased drain-source dv/dt and decreased power dissipation during each switching period. The results of measurements show that the effect of increased dV_{d-s}/dt is to broaden the E-field spectrum and to increase the amplitude of the field harmonics by up to 30dB, as seen by comparing Fig. **24** and **25**.

The E-field radiation measurements taken at distances further than 1m from the SMPS both with and without a heatsink do not show any difference. Similar results are observed from the FDTD simulation results for the 200 MHz range and 3m observation distance where the heatsink does not increase radiated fields significantly.

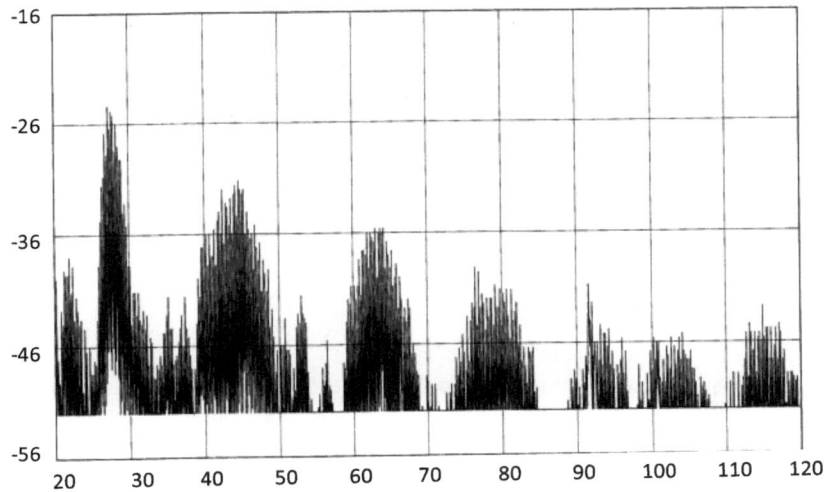

Figure 25: E-field signals of the SMPS with the increased dV_{d-s}/dt, [dBmW/ MHz], [19].

CONCLUSIONS

The key demands made by PCB mounted power circuits are related to low conduction losses, improved thermal performance, and lower inductance board layouts. Thermal management of power devices involves the use of heatsinks which are seen as parasitic elements from an EMI point of view. This study provides suggestions to model heatsinks and demonstrates how this problem can be solved using numerical electromagnetic computation methods.

Common mode coupling mechanism is the main component of the analytical and numerical analysis in this study. The analytical solution is limited to simple PEC geometries typically used in electrostatics studies. The electrodynamic character of the parasitic effects and the physical structure of the power devices can only be completely understood by using a three dimensional numerical simulation, where all the electric and magnetic fields can be calculated through a volume containing the heatsink and power device. Knowledge of the stray capacitance is important for predicting conducted EMI and making comparisons between EMI effects of different heatsink designs.

EMI diagnostic comparison measurements are a rapid way of obtaining useful information about the possible effects of heatsinks on near field radiation and can also be used to validate FDTD heatsink models.

The results of the presented work do not detail the precise levels of radiated fields; all the results are based on relative comparisons. Precise levels of radiated fields cannot be easily compared with the numerical modeling results because of the dependence on the excitation sources. The results of modeling show that the intensity of the field radiation does not increase proportionally with frequency and at some frequencies the heatsink becomes a more effective radiator. This is not possible to predict by using simplified analytical expressions; numerical modeling provides more accurate solutions. However, numerical modeling is also limited since the modeling results depend on the representation of the physical properties of the excitation sources. The results of the FDTD computations of the heats sink effects were obtained with two different power semiconductor packages. If the common mode EMI approach is used when modeling the excitation source in FDTD computation then it can be concluded that the heatsink increases radiated EMI.

The experiments were carried out using a low power SMPS with a switching frequency above 100 kHz. Based on the results obtained it can be concluded that a typical SMPS circuit radiates electric field harmonics within a frequency range of 200MHz in close vicinity of the SMPS. The heatsink generally enhances the electric field amplitude spectrum at certain frequency ranges. However, in some cases the heatsink appears to have the opposite effect and can decrease the amplitude of radiated signals at frequencies above a certain range. An increase in voltage variation and switching speed has considerable impact on the generation of radiated E-field signals. This confirms the assumption that the switching device and its voltage variation are the main source of common mode based EMI for an SMPS. The effect of the heatsink on the electric field signals measured at a distance of 1m was too small to be measured for frequencies below 200 MHz. This confirms the predictions obtained from antenna models of heatsink that the far field radiation enhancement due to the heatsink is not significant at these frequencies.

REFERENCES

[1]   Mardiguian M. Controlling radiated emissions by design. 6th ed. Kluwer Academic Publishers: Dordrecht 1999.
[2]   http://www.infineon.com/cms/
[3]   Sawle A, Blake C, Mariae D. Novel Power MOSFET packaging technology doubles power density in synchronous buck converters for next generation Microprocessors. International Rectifier, APEC, 2002.
[4]   Tihanyi L. Electromagnetic compatibility in power electronics. IEEE Press: 1995.
[5]   Kunz KS, Luebbers RJ. The finite difference time domain method for electromagnetic. CRC, Boca Raton 1993.
[6]   Bosanac T. Teoretska elektrotehnika 1. Tehnicka Knjiga: Zagreb 1970.
[7]   Kreyszig E. Advanced engineering mathematics. 9th ed. Wiley: Singapore 2006.
[8]   Stellari F, Lacita AI. New formulas of interconnect capacitances based on results of conformal mapping method. IEEE Trans Elec Dev 2000; 47(1).
[9]   Yee KS. Numerical solution of initial boundary value problems involving Maxwell's equations in isotropic media. IEEE Trans Ant Prop 1966; 14: 302-307.
[10]   Tafflove A. Computational electrodynamics: The finite difference time domain method. Artech House: Boston 1995.
[11]   Courant R, Friedrichs K, Lewy H. On the partial difference equations of mathematical physics. IBM J Res and Develop 1967; 11(2): 215-234.
[12]   Liao ZP, *et al*. A transmitting boundary for transient wave analysis. Scientia Sinica1984; series A; XXVII; 10.
[13]   Balanis C. Antenna theory, analysis and design. 2nd ed. Wiley: 1997.
[14]   Kraus JD Antennas. McGraw Hill: 1998.
[15]   APLAC Circuit simulation and design tool, Aplac Solutions Corporation 1999.
[16]   Felic G, Evans R. FDTD based analysis of heatsink effects in SMPS circuits. In: European power electronics conference, DS3.8; Graz, Austria; 2001; pp. 1-8.
[17]   Consoli A, Musumeci S, Oriti G. An innovative EMI reduction design technique in power converters. IEEE Trans EMC 1996; 38(4): 567-575.
[18]   Igarashi S, *et al*. Analysis and reduction of radiated EMI noise from converter systems. Electrical Eng Jpn 2000; 130(1): 106-117.
[19]   Felic G, Evans R. Study of heatsink EMI effects in SMPS circuits. IEEE International Symposium on EMC; Montreal, Canada 2001; 11: 254-259.

CHAPTER 4

Motor Cable Influence on the Conducted EMI Emission of the Converter Fed AC Motor Drive

Jarosław Łuszcz[*]

Faculty of Electrical and Control Engineering, Gdańsk University of Technology, G. Narutowicza 11/12, 80-952 Gdańsk, Poland

Abstract: Investigation of conducted electromagnetic interference in AC motor drives fed by pulse width modulated voltage converters requires considering parasitic capacitances in converters, motor windings and feeding cables to be taken into account. Motor voltage transients and related conducted electromagnetic emission are significantly correlated with resonance effects occurring in load circuits. The levels of intensity of these phenomena depend mostly on frequency dependant impedance - frequency characteristics of motor windings with an accompanying feeding cable. An analysis of frequency converter load impedance characteristics allows for identification and determination of representative frequency ranges in which the foremost contributions to EMI noise generation have voltage ringing phenomena associated with the load parasitic capacitances. This chapter presents a method to model an AC motor with a feeding cable. AC motor windings have distributed parasitic capacitances and a particular focus on the influence of the feeding cable's parameters is modeled as a ladder circuit model. The proposed circuit model allows for an analysis of the influence of the cable parameters on conducted EMI emission generated by an AC motor drive system. The simulation results based on the proposed ladder circuit model are verified by the experimental tests which were carried out for an exemplary adjustable speed AC motor drive application with different lengths of feeding cables.

Key Words: Bearing current, Broadband model, Cable shield, Common mode impedance, Common mode transfer impedance, Common mode voltage, Dielectric permeability, Differential mode current, Distributed model, EMI propagation, Grid impedance, Impedance mismatch, Line impedance stabilization network, Ladder circuit model, Load impedance, Lossy transmission line model, Lumped model, Modulation carrier frequency, Motor feeding cable, Motor windings, Power grid impedance, Propagation velocity, Reflection coefficient, Reflection effects, Resonance effects, Resonance frequencies, Self impedance, Shaft voltage, Shielded cable, Smith chart, Standing waves, Wideband modelling, Transmission line effects, Transition frequency, Transmission line.

INTRODUCTION

Pulse width modulated (PWM) voltage source inverters commonly applied in contemporary adjustable speed drives (ASD) are a cause of accompanying undesirable high frequency (HF) side effects in the powered AC motor and supplying grid [1-6]. Increase of pulse width modulation carrier frequency and decrease of converter transistors' switching time intensify existing HF problems in various ways. One of the foremost problems related to these phenomena is generation of high frequency stray current pulses flowing through drive components - due to high levels of output voltage steeps (dv/dt) - and unavoidable parasitic capacitances [7-9]. There are two components of these currents: a differential mode (DM) current which flow between power lines and a common mode (CM) current flowing between all energized components and the ground line.

HF components of stray currents, flowing in all conducting components of the ASD due to the existence of parasitic inter-capacitances, are not limited directly by standard regulations related to power electronic converters. Nevertheless the foremost consequences of the stray HF current flow, especially through the parasitic capacitances, is the generation of HF voltage components appearing in all conductive elements of the ASD system. These HF voltage components, especially CM voltages, are highly hazardous and are the fundamental source of EMI conducted emission noise of power electronic converters.

Detailed modelling of an ASD system in a wide frequency range to analyze all interactions between a PWM voltage converter and an AC motor with cable is challenging and requires the use of complex methods and broadband models [10-13]. One of the main problems to model inverter feed AC motor drives in a wide frequency range is to accurately

---

***Address correspondence to Jarosław Łuszcz:** Gdansk University of Technology, Sobieskiego 7, 80-216 Gdańsk; Tel: +48 58 3472534; E-mail: jlusz@ely.pg.gda.pl

Firuz Zare (Ed)

identify the parasitic parameters of all drive system components, especially in HF range. For the conducted EMI frequency range usually up to *30 MHz* the predominant model identification problem is related to parasitic capacitances of the system components. These capacitances induce various HF objectionable effects such as: local resonances in parasitic sub-circuits and increased flow of harmful CM currents affiliated to them [3, 5, 14-18].

Elevated CM currents generated on the output side of the frequency converter due to the extraordinary resonance effects in the feeding cable and the AC motor windings, propagate to all the surrounding circuits and return back to the supply grid through the frequency converter. The effectiveness of this propagation path is strongly related to the impedance of grounding and bonding connections of the converter in the HF range. Nevertheless, it is extremely difficult to achieve very high quality of converter's grounding and bonding connections, so high levels of CM currents generated in the converter's output-site can easily and efficiently approach the line-side of the converter and finally the supplying grid.

This chapter presents a method for analysing the influence of high frequency parameters of AC motor windings with a feeding cable on the global conducted EMI emission levels injected into the supplying power grid. The proposed identification method of the load broadband circuit model parameters is based on the load impedance frequency characteristics analysis. Determination of the most meaningful resonance frequencies observed on the load impedance characteristic allows for a ladder circuit model configuration arrangement with the appropriate number of rungs representing most meaningful resonances. Using this method the complexity of the circuit model can be kept within a reasonable complexity level, which is essential for its parameter identification process.

The experimental results prove that the high frequency phenomena in the evaluated system are strongly correlated in an adequate frequency range with the converter load impedance frequency characteristic. Finally, based on the simulation and experimental results it can be underlined that general EMC performance of an ASD with relation to the supplying power grid is noticeably dependant on the converter load HF parameters. Significant improvement in levering conducted EMI emission of an ASD can be obtained by selection of an adequate motor cable and its proper installation to avoid undesirable CM currents.

EMI SOURCES IN CONVERTER FED AC MOTOR DRIVES

Power electronic converters with fast switches as IGBTs or MOSFETs widely used in AC motor drives are intrinsic sources of high level EMI in a broad spectrum range. The main source of the conducted and radiated EMI generation is due to rapid voltage and current changes caused by each switching process of semiconductor power devices. Signals with high levels of dv/dt and di/dt can be efficiently coupled by parasitic capacitances and mutual inductances with nearby objects, which make possible propagation of conducted and radiated emission noise outside a converter into surrounding electromagnetic environment. Classically spectrum of time domain signals can be calculated by using Fourier analysis. Nevertheless, it is enormously time consuming method which leads to relatively complex results even with using many tolerable simplifications. Complex equations achieved by this method can not clearly show the influence of fundamental signal parameters on its spectrum. More efficient approach for estimating spectrum of typical signals occurring in power electronic converters is graphical method of signal spectrum envelope estimation. Graphical methods are based on the assumption that complex signals can be decomposed into elementary simple signals for which the asymptotical envelope of the spectrum is relatively simple to determine.

Particular voltage waveforms happening typically in power electronic converters can be considered, with some simplification, as trapezoidal (Fig. **1**), where V_{DC} is a switched voltage, $f_0=1/T$ is the switching frequency, $d=t_{on}/T$ is the duty cycle, t_r and t_f are voltage rising and falling times, respectively.

Presented voltage waveform V(t) can be decomposed into two ideal rectangular pulses: the first one with magnitude VDC and duration ton and second with magnitude VDC/tr or VDC/tf and duration tr or tf which is equivalent to voltage derivative dv/dt.

Spectrum of ideal rectangular waveform can be determined by using *sin(x)/x* function; therefore the spectrum of trapezoidal signal as a convolution product of two ideal rectangular waveforms can be calculated using formula (1).

$$V_{(n \cdot f_0)} = 2d \cdot V_{DC} \left| \frac{\sin(n\pi d)}{n\pi d} \right| \left| \frac{\sin(n\pi t_r f_0)}{n\pi t_r f_0} \right| \qquad \textbf{(1)}$$

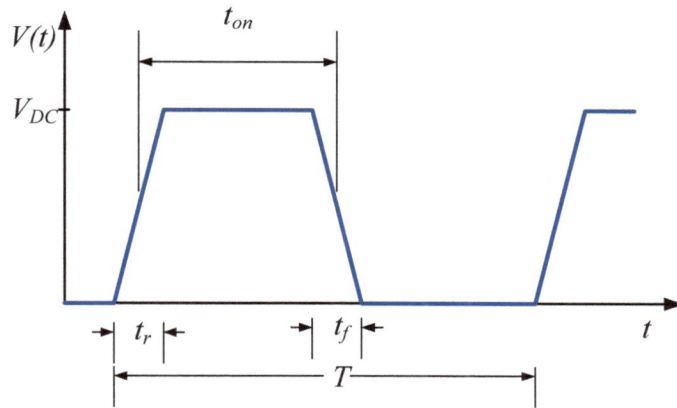

Figure 1: Simplified output voltage waveform of frequency inverter: V_{DC} - switched DC voltage, T - switching period, t_{on} - voltage pulse duration, t_r - voltage rising time, t_f - voltage falling time.

Two examples of voltage spectra calculated by this method for two simplified voltage waveforms with evidently different and typically occurring in contemporary ASD parameters sets are presented in Fig. **2**. On the presented spectra characteristic frequencies and its multiples correlated exactly with time constants of analysed waveforms t_{on} and $t_r=t_f$ can be observed. For each of these frequencies according to formula (1) spectra magnitudes have zero values and between them obtain locally highest values.

The spectrum of trapezoidal waveform obtained using this method can be furthermore simplified by defining spectrum envelope's characteristic transition frequencies $f_{T1}=1/\pi\,t_{on}$ linked to voltage pulse duration t_{on} and $f_{T2}=1/\pi\,t_r$ linked to voltage pulse rise t_r and fall time t_f. For frequencies below f_{T1} the asymptotical envelope is constant, then for higher frequencies it decreases with the slope of *-20 dB/dec*. The influence of the rising and falling edges of the switched voltage can be clearly seen on the spectrum envelope characteristic as a next transition frequencies f_{T2}, above which the spectrum magnitude decrease twice stronger with the slope of *-40 dB/dec* (Fig. **3**).

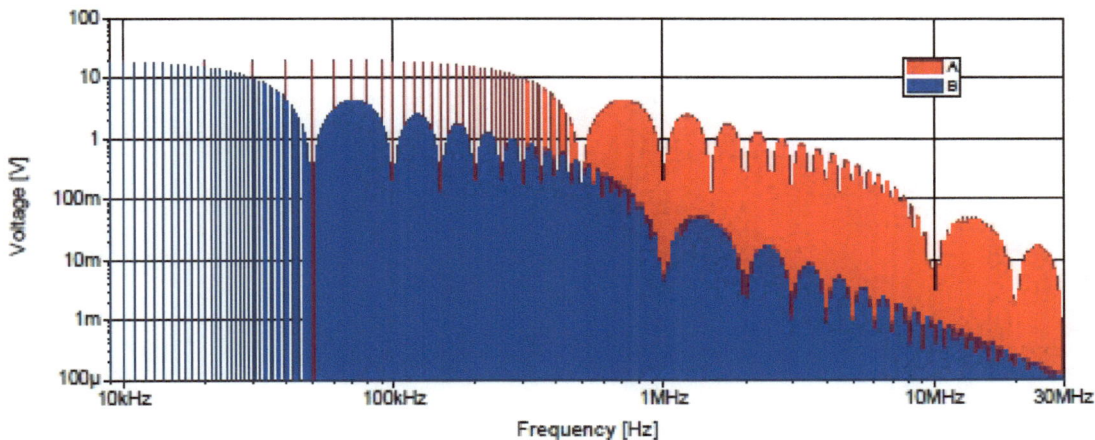

Figure 2: Spectra examples of trapezoidal voltage signals: spectrum A - for $V_{DC} = 500\ V$, $f_0 = 10\ kHz$, $d=2\%$, $t_r = t_f = 100\ ns$; spectrum B - for $V_{DC} = 500\ V$, $f_0 = 1\ kHz$, $d=2\%$, $t_r = t_f = 1\ \mu s$.

PROPAGATION OF EMI CURRENTS IN ASD DRIVES

A classical ASD converter includes a line side AC/DC converter feeding an internal DC voltage bus and a load side DC/AC inverter fed by this DC bus. Analysis of EMI propagation in the ASD is mostly related to the load side converter, which is usually voltage source PWM controlled inverter with IGBT switches. In this type of inverter, the output voltage stress (*dv/dt*) due to switching transient is usually significant due to the bipolar DC bus voltage commutation during transistor switching time. Nevertheless, contemporary line side PWM controlled converters can also have significant effect on the EMI generation but in this chapter we only analyze the ASD with a line side diode rectifier. This simplification allows observing more clearly load side inverter effects on the evaluated phenomena by minimizing simultaneous influence of line side converter.

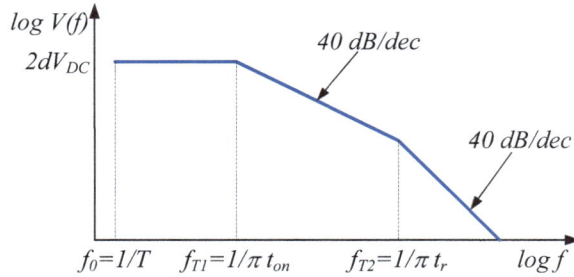

Figure 3: Asymptotical spectrum envelope of trapezoidal voltage waveform (*T* - waveform period, *d=t$_{on}$/T* - duty cycle, *t$_r$*, *t$_f$* - voltage rise and fall time.

The ASD configuration presented in Fig. **4** has two primary CM current paths: the first one *I$_{CM1}$* related to the line side converter, the supplying line, the power grid impedance and the grounding impedance and the second one *I$_{CM2}$* related to the load side inverter, the motor feeding cable, the motor windings, the motor impedance and the grounding impedance.

Analysis of the CM currents flow in this circuit is difficult because of difficulty to identify impedance-frequency characteristics of the circuits, especially the power grid impedance and the converter load impedance within the analysed frequency range.

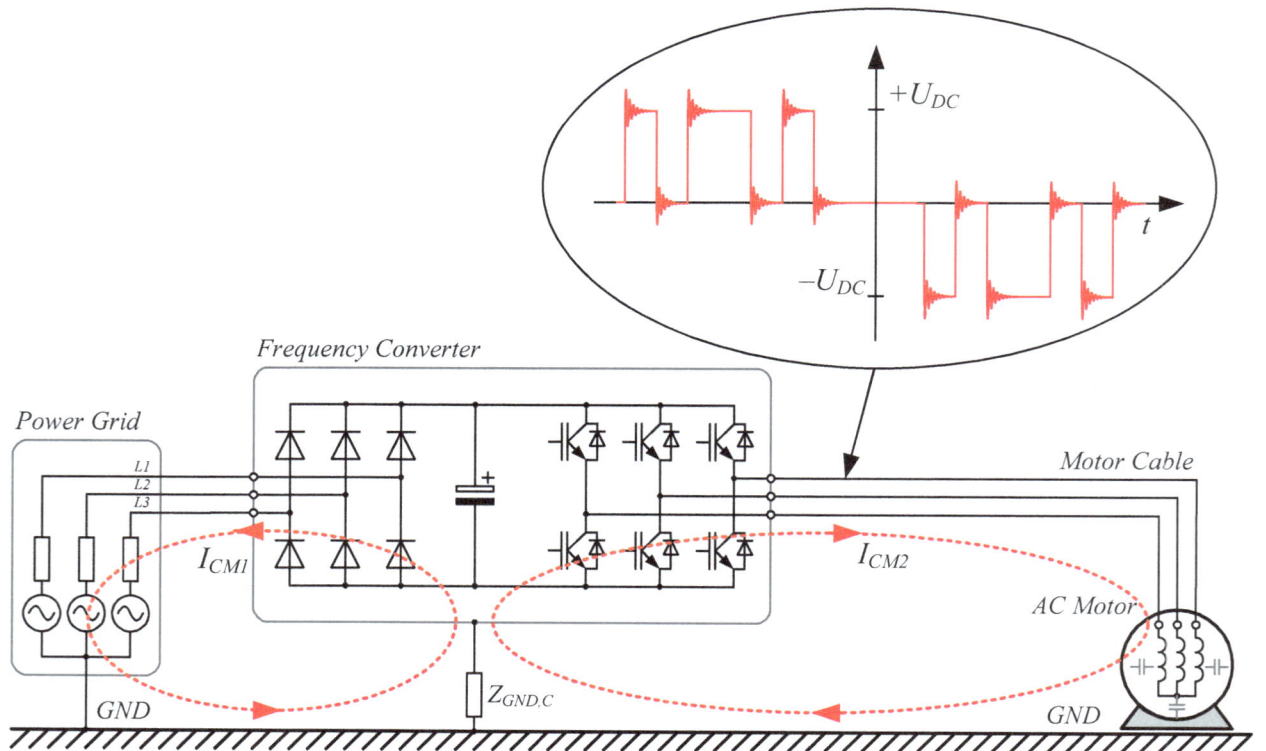

Figure 4: Two primary CM current paths in the adjustable speed AC motor driver.

Supplying power grid impedance for the conducted EMI frequency range can be estimated approximately based on the measurement of system reaction to injected high frequency currents in the particular point; however it varies permanently due to power system load changes. Difficulties with determination of inverters load CM impedance are associated with parasitic capacitances of the motor windings and the feeding cable with respect to the ground. The parasitic capacitances are mostly essential for the load CM impedance variations. Distributed character of the parasitic capacitances jointly with rapid voltage changes induces transmission line effects in the motor feeding cable and its windings. Commonly occurring frequency dependant impedance mismatch between the inverter outputs, the feeding cable and the motor windings complicates the analysis. The high frequency conducted EMI currents flow through the cable and the motor can be clarified by DM and CM current components as shown in Fig. **5**.

The magnitude of CM and DM conducted EMI currents and its allocation along the feeding cable and the motor windings depend foremost on the parasitic capacitances and voltage steepness dv/dt. The DM currents generally circulate through the load impedances with indirect influence on the conducted EMI emission noise. Nevertheless any asymmetries in a three phase system can generate more CM currents components. The CM currents circulate through the motor windings and its feeding cable and it is coupled galvanically with the line side of the converter and the grounding impedance $Z_{GND,C}$ which allows for propagation of output side EMI emission towards the power grid.

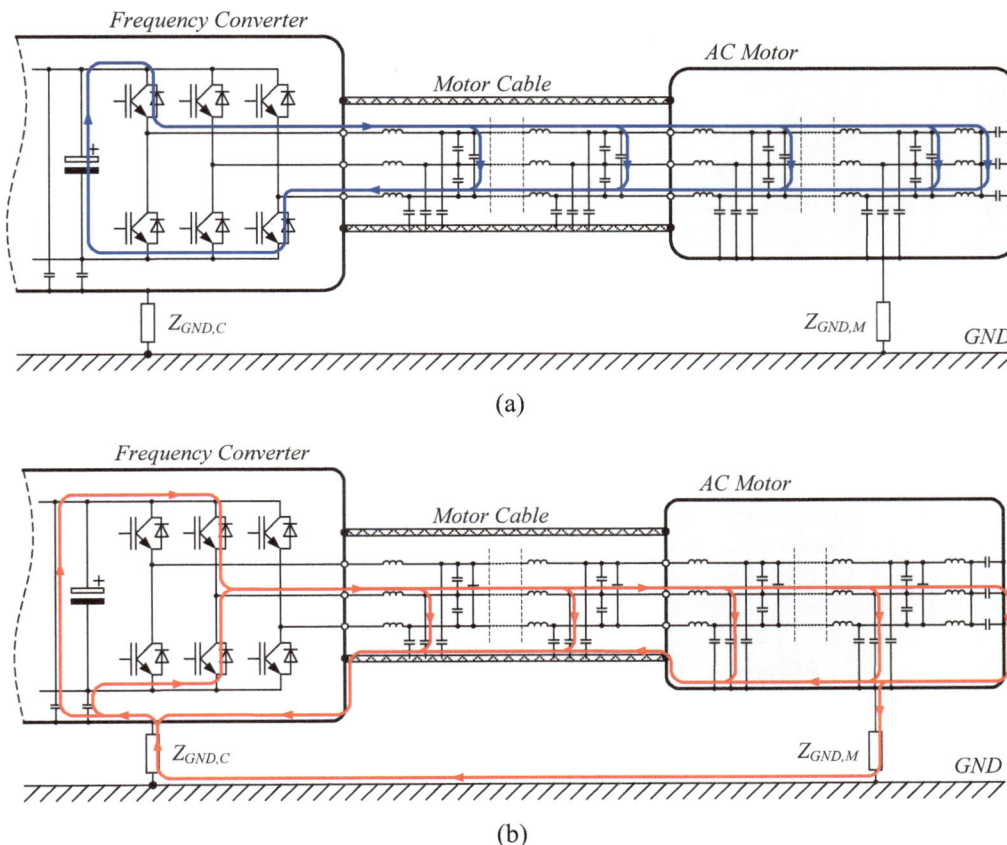

Figure 5: EMI current path in ASD output side: (**a**) - differential mode, (**b**) - common mode.

WIDEBAND MODELLING OF THE CONVERTER LOAD

Modelling of AC Motor Windings

Detailed wideband modelling of an AC motor with a feeding cable fed by a frequency converter is rather difficult because of distributed parasitic capacitances which should be determined. Complex and unequal distribution of the parasitic capacitances along the windings make the identification process extremely difficult. For a typical AC motors with a rated power of hundreds of kW in the frequency range above several kHz usually distributed parasitic capacitances became significant and cannot be omitted. An effortlessness of the identification of the parameters of the AC motor windings model is the main motivation for developing rationally simplified models.

A simplified model of the distributed capacitances in AC motor windings and feeding cables in a wide frequency range can be defined based on a ladder circuit model with an appropriate number of rungs - adequate for the given frequency range and expected accuracy. This simplification can be effectively used in a frequency range in which lumped models are not accurate enough and detailed determination of distributed parasitic parameters of the evaluated system is too complicated. General configuration of the proposed circuit model with the N - step ladder circuit is presented in Fig. **6**.

The CM transfer impedance as a relation between the winding voltage at the motor terminal A and the total CM currents flowing through the distributed winding to the ground capacitances can be formulated based on the circuit model presented in Fig. **6** as follows:

$$Z_{T(CM)}(s) = \frac{u_A(s)}{\sum_{k=0}^{N} i_{C_{gk}}(s)} \tag{2}$$

Assuming that the winding self capacitances C_{s1}, C_{s2}, ..., C_{sN} are usually considerably smaller then the winding to ground capacitances C_{g0}, C_{g1},..., C_{gN}, the evaluated transfer impedance can be determined analytically based on the proposed circuit model with using formula (3).

$$Z_{T(CM)}(s) \approx \cfrac{1}{sC_{g0} + \cfrac{1}{sL_1 + \cfrac{1}{sC_{g1} + ... + \cfrac{1}{sL_N + sC_{gN}}}}} \tag{3}$$

Example of the simplified CM transfer impedance characteristic determined based on the ladder circuit model analysis is presented in Fig. **7**. This characteristic exhibits a number of specific resonance frequencies associated with parallel resonances f_{p1}, f_{p2}, f_{p3}, ..., f_{pN} and serial resonances f_{s1}, f_{s2}, f_{s3}, ..., f_{sN}. These resonances are correlated with poles and zeros of the proposed transfer impedance function $Z_{T(CM)}$. The CM impedance characteristic segments between the mentioned resonance frequencies are related to the equivalent lumped parasitic winding capacitances and adequate quantity of inductances denoted as L_1, L_2, ... L_N that form the winding inductance.

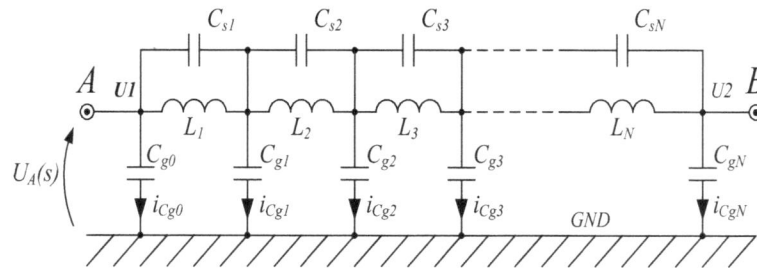

Figure 6: Distributed parasitic capacitances between motor's windings and ground, simplified lumped representation by N-step LC ladder circuit.

Modelling of a Feeding Cable

A standard AC Motor feeding cable can be represented as three conductor lines referenced to the ground and in a shielded cable which are commonly used for feeding AC motors in ASD, the cable shield is represented the ground. The cable shield also equalizes conductor to ground parasitic capacitances and a fully symmetric circuit model can be implemented as shown in Fig. **8**.

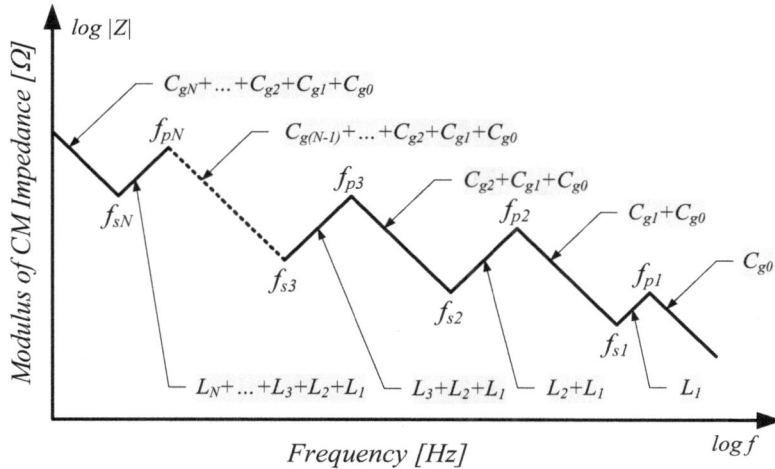

Figure 7: Simplified theoretical impedance-frequency characteristic of CM transfer impedance of N - step ladder circuit model.

Length and impedance of the AC motor feeding cable are essential for the ASD general performance and its particular influence on the conducted and radiated electromagnetic emissions. Commonly used AC motor cables are from few meters up to few hundreds of meters long and are divided into two classes: short and long cables. In many ASD application and technical notes a critical cable length is defined which means that for the feeding cable longer than the critical value adverse effects are expected and some protective methods should be applied. The critical length of the cable which allows classifying as a short or as a long cable is difficult to be determined exactly and it also depends on spectral characteristics of transmitted signals and pulse propagation velocity. Therefore the critical cable length can be referenced to the wave length of transmitted signal.

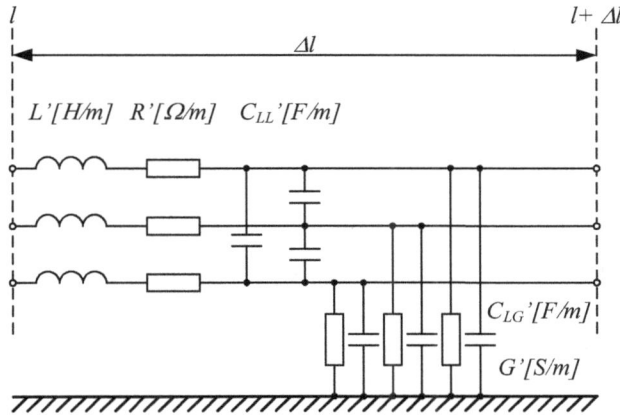

Figure 8: Classic lossy transmission line circuit model of AC Motor feeding cable.

The contemporary ASDs generate broader voltage and current spectrums exceeding tens of MHz because of the common tendency to decrease transistor switching time and increase modulation carrier frequency to improve power conversion efficiency. From this point of view it can be said that short cables are very rare, because ASD's voltage output spectrum is usually broad enough to generate standing waves even in a cable with few meters in length. Nevertheless, the power spectrum density of the generated signals in a high frequency range is decreased noticeably and lossiness of the cable is increased with frequency. Thus the signal reflection effects in high frequency range are noticeably reduced. The critical cable length also depends on the defined levels of reflection effects which can be tolerated.

Theoretically, a critical cable length parameter, which allows estimating the intensity level of exposure of the ASD to reflection phenomena in the AC motor cable is commonly estimated based on the converter output voltage rise time t_r. The wave length related to voltage rise time can be calculated by formula (4) where v is a signal propagation velocity in the evaluated cable.

$$\lambda = t_r \cdot v \tag{4}$$

The signal propagation velocity in typical power cables is related to the relative dielectric permeability of cable insulation ε_r (5) and also correlated with per meter parameters: resistance R' [Ω/m], inductance L' [H/m], capacitances C'_{LL}, C'_{LG} [F/m] and conductance G' [S/m] as by propagation factor k_p according to formula (6) where inductance L_0 and capacitance C_0 are equivalent per unit length parameters [19].

$$v = \frac{1}{\sqrt{L_0 C_0}} = \underbrace{\frac{1}{\sqrt{\mu_0 \varepsilon_0}}}_{c} \cdot \frac{1}{\sqrt{\varepsilon_r}} \tag{5}$$

$$k_p(s) = \sqrt{(R_0 + sL_0)(G_0 + sC_0)} \tag{6}$$

Generally the signal propagation velocity in the power cables is approximately two times smaller then light speed c, because dielectric permeability for typical cables insulation varies from 3 to 8, so the obtained velocity is usually within the range of 40~60 % of the light speed. The characteristic cable impedance Z_C Eq. 7 allows determining the real reflection coefficient Γ at the motor terminals using Eq. 8 where Z_M is the motor winding input impedance.

$$Z_C(s) = \sqrt{\frac{(R_0 + sL_0)}{(G_0 + sC_0)}} \approx \sqrt{\frac{L_0}{C_0}} \tag{7}$$

$$\Gamma(s) = \frac{Z_M(s) - Z_C(s)}{Z_M(s) + Z_C(s)} \tag{8}$$

For the given frequency of transmitted signal, the critical length of the cable can be correlated to the wave length determined by consideration of the real velocity of the signal propagation. The minimum length of the cable for which the reflection effects can be noticeably increased is $\lambda/4$.

Parameters per length of presented type of transmission line model for particular cable can be determined analytically or estimated experimentally by measuring the CM impedance in open circuit Z_{OC} and short circuit Z_{SC} configurations. Based on the measured results, a characteristic transition frequency f_T can be defined as a point where measured impedance in the open circuit Z_{OC} and the short circuit Z_{SC} are equal. This transition frequency divides the evaluated frequency range into sub ranges. In the frequency range below transition frequency f_T, lumped models can be efficiently used for analysis, but for the higher frequencies lossy transmission line models of the cable are required. Example of measurements results for the three conductor *100m* long shielded cable is presented in Fig. **9**. In the presented example, for frequency range above *100 kHz* the distributed model is unavoidable for adequate modelling of the cable behaviour in the whole conducted EMI frequency range up to *30 MHz*. The value of Z_{OC} and Z_{SC} impedance at the transition frequency is the same and equal to characteristic impedance Z_0 of the evaluated cable.

Wideband Identification of an AC Motor with a Feeding Cable

The presented investigation of the conducted EMI emission in the evaluated AC motor drive system has been focused on the CM currents. The influence of various interactions between the CM and the DM emission noise existing in test system - which are extremely difficult to eliminate entirely - have been minimized by keeping values of all significant parameters as identical as possible. The only difference is the length of the motor feeding cable. The common mode impedance between the load terminals and the ground in the conducted EMI frequency range has been measured using a vector network analyser. The measurement results obtained for the evaluated AC motor with the short and long cables are presented in Fig. **10**.

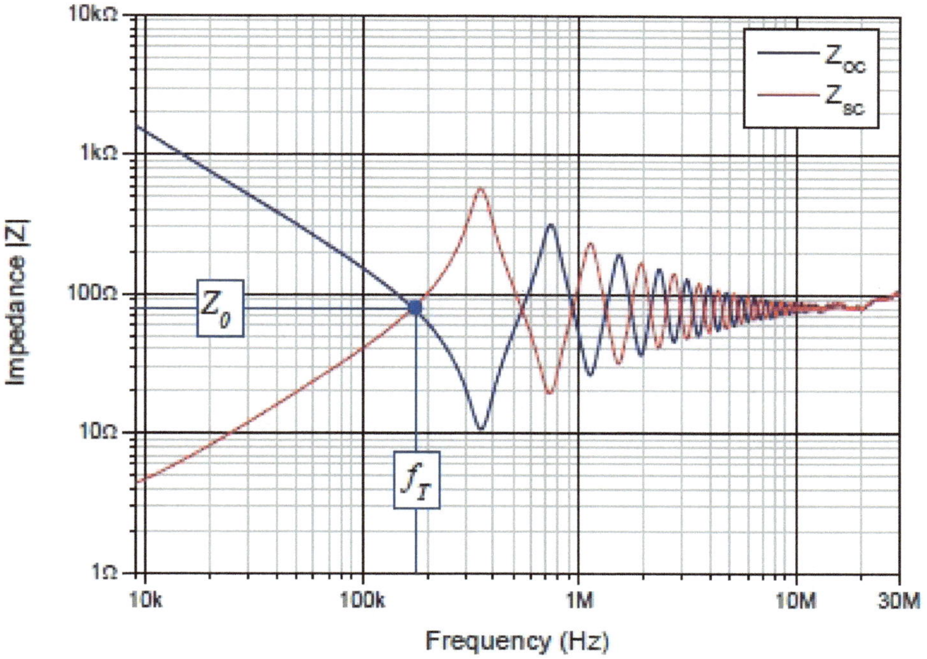

Figure 9: Open circuit (OC) and short circuit (SC) CM impedance of 100m long motor cable.

The Smith chart representation of the measured impedances clearly shows main differences in the load performance according to the CISPR B frequency range *150 kHz - 30 MHz*. The number of resonances, correlated particularly with the long feeding cable as a transmission line which is not matched to the motor input impedance is evidently visible as extra loops on the Smith chart. As a reference analogous characteristic of the motor with the short feeding cable is also shown. The most meaningful resonance frequencies which are determined based on the measured impedance characteristic are *4 MHz* for the AC motor winding with a short cable and *0.9 MHz*, *2 MHz*, *3 MHz*, *6 MHz* for the AC motor winding with a long cable which are hardly correlated with the load parameters.

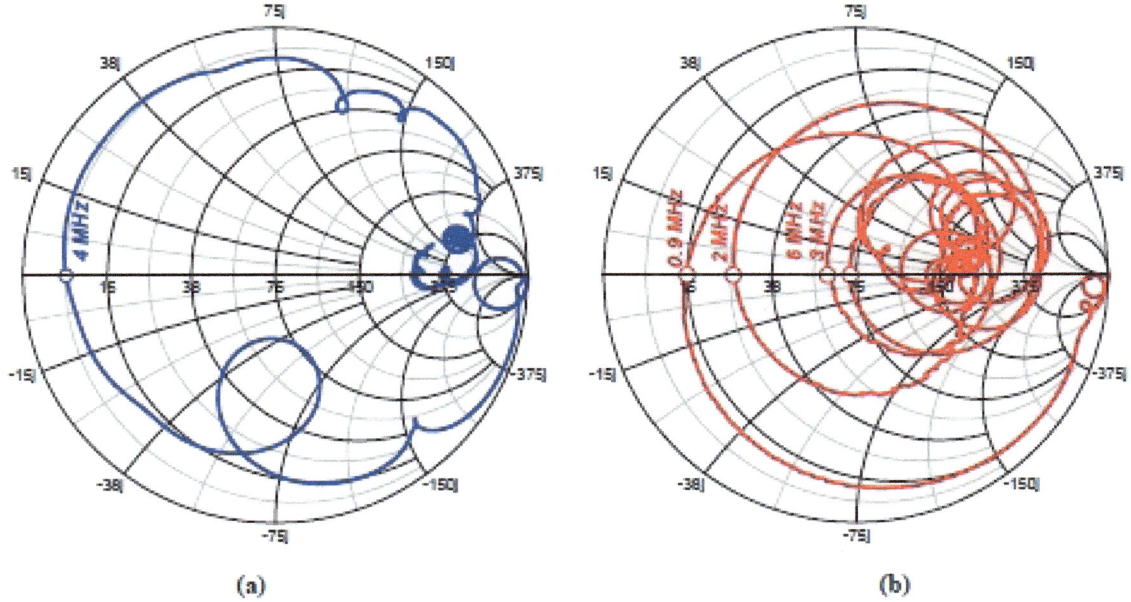

(a)   (b)

Figure 10: Common mode impedance of the evaluated AC motor with short (**a**) and long (**b**) feeding cable.

MOTOR CABLE AS A SOURCE OF ADDITIONAL CONDUCTED EMI EMISSION

High Frequency Interactions between an AC Motor and a Feeding Cable

An adequate circuit model of an AC motor with feeding cable is presented in Fig. **11**, where the AC motor is represented as an N - rung LC ladder circuit whereas the feeding cable is represented as a lossy transmission line.

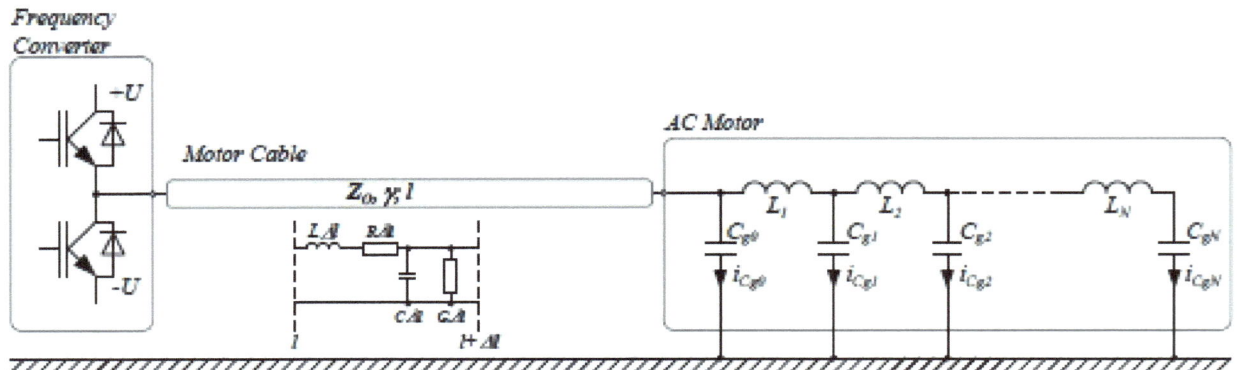

Figure 11: Circuit model of AC motor winding with feeding cable for broadband CM currents analysis - one phase representation.

The foremost benefit of the circuit model is its relatively low complexity level in relation to the achieved accuracy. A reasonable balance kept between a model simplicity and its adequacy allows for significant reduction of its parameters identification efforts and decrease radically a computational overheads of the simulation analysis.

An identification of parameters of the circuit model of the AC motor with feeding cable can be completed based on the terminal impedance measurement. Based on this model various simulations have been carried out using the motor and the feeding cable parameters which are identified experimentally. Samples of simulation results obtained for the tested AC motor with two lengths of the feeding cable are presented in Fig. 12.

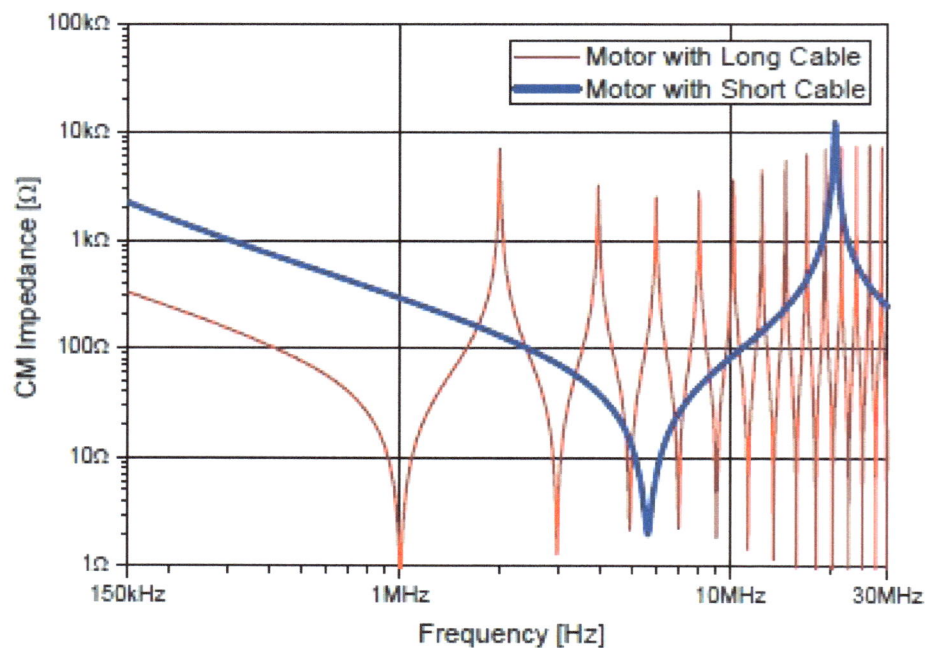

Figure 12: CM impedance of AC motor and AC motor with cable - simulation results based on the proposed circuit model.

Presented simulation results allow studying theoretical influence of the feeding cable on the CM impedance. Especially the specific resonance frequencies can be clearly correlated with the length of the AC motor feeding cable. On the basis of these simulation results, the smallest resonance frequency of the converter load can be estimated for various lengths of the feeding cables. The minimum load impedance observed for this smallest resonance frequency cause the maximum CM current.

Experimental Test Bench for CM Currents Measurement

Experimental evaluation of the load influence on conducted EMI emission of an AC motor fed by a frequency converter has been done for a typical low power converter with a diode rectifier at the input side. Rated powers of the tested AC motor and the converter are *7.5 kW*. The ASD frequency converter with a three phase diode rectifier as a line side converter has been selected to minimize undesirable high frequency interaction between the line side and the load side of the ASD converter. Therefore in the tested ASD drive, the main active source of the high frequency emission in a frequency range up to several MHz are power switches, controlled by PWM modulated switching signals.

General configuration of the experimental test bench is presented in Fig. **13**. In the presented test bench, the line side and the load side CM currents have been measured. The first one is associated with the input CM current path I_{CM1}; the second one is associated with the converter output CM current path I_{CM2}. Both of these CM current loops are coupled by the common equivalent impedance $Z_{GND,C}$ which represents capacitive and inductive couplings of the converter with respect to the ground. A Line Impedance Stabilization Network (LISN) has been used for minimizing power grid impedance influence on the measurement results. The CM currents have been measured using two wide band current probes with equalized transfer characteristics to improve measurement accuracy. A four pole, *50 Hz*, AC motor with no mechanical load, is connected to the converter's output terminals by a three wire shielded cable. Two types of feeding cables with different lengths are used: the first one has *0.2 m* long and the second one has *20 m* long. In both cases a same ground connection (about *0.5 m*) between the converter and the motor is made as a separate wire with minimized self impedance.

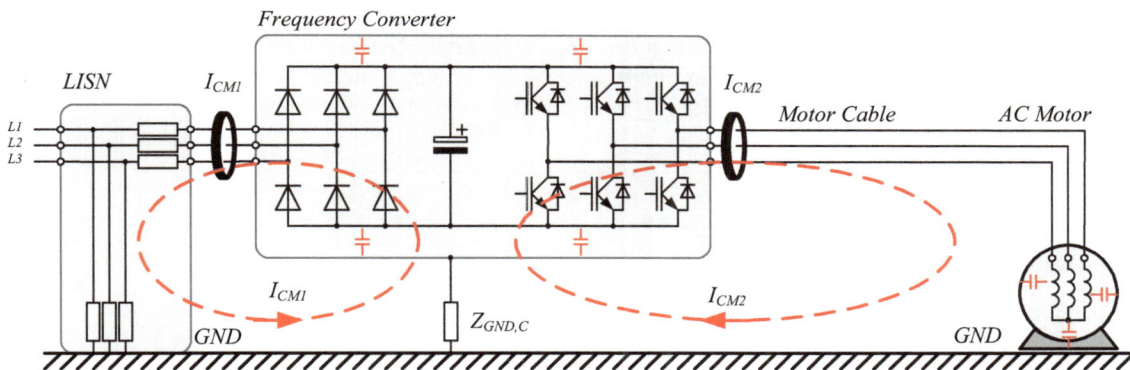

Figure 13: Experimental test bench for CM currents measurement.

Experimental Investigation of Motor Cable Influence on EMI Emission

An analysis of correlation of conducted EMI emission level with the load CM impedance characteristic can be carried out in time and frequency domains. Analysis in time domain is problematic because transistor switching conditions change noticeably and periodically within relatively long periods of AC drive output waveform and power grid. The main variations are related to commutated currents because their averaged values change periodically, nearly sinusoidally according to modulated AC drive output frequency. The switched voltages also vary due to DC link voltage fluctuations caused by supplying rectifier pulsations and load current changes. These changes cause that each following switching process is different. Effects of these transistor switching condition variations are clearly visible as a noticeable differences between each CM current pulse. An example of the input CM current measured for the AC motor connected to the converter with the short cable is presented in Fig. **14**, which has two specific ringing frequencies.

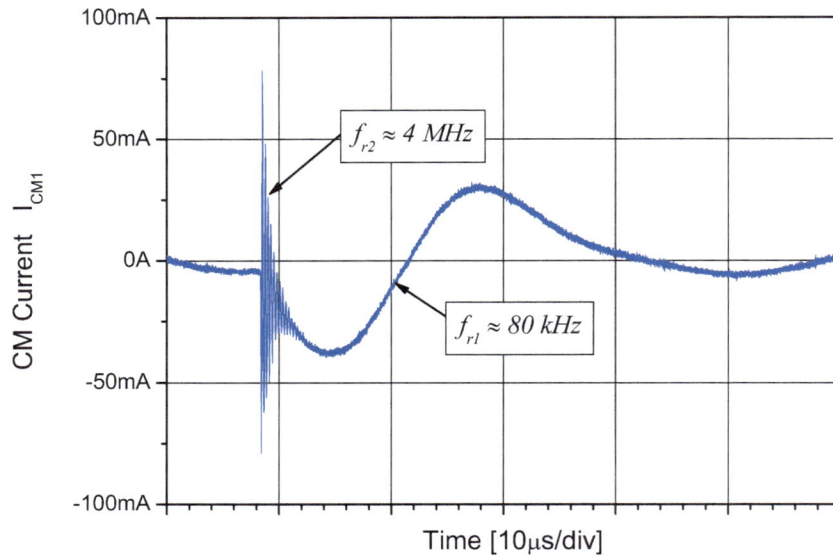

Figure 14: An example of measured single pulse of CM input current I_{CM1} of the evaluated converter with AC motor with short cable.

One of the ringing harmonic components is related to the input CM current I_{CM1} (Fig. **13**) with relatively low frequency (about *80 kHz*) and the second ringing harmonic component is related to the load side CM current, I_{CM2} with relatively high frequency, approximately *4 MHz*.

A comparison between the magnitude comparison of the measured CM current pulses I_{CM1} and I_{CM2} shows that the low frequency component is usually less significant in the load side loop (Fig. **15**) and the high frequency component is more significant in the load side loop Fig. **16**. This relation between the CM current components in the line and the load sides proves that the grounding impedance $Z_{GND,C}$ which is mutual for both CM current loops is the most meaningful coupling path between the line and load sides of the CM current loops.

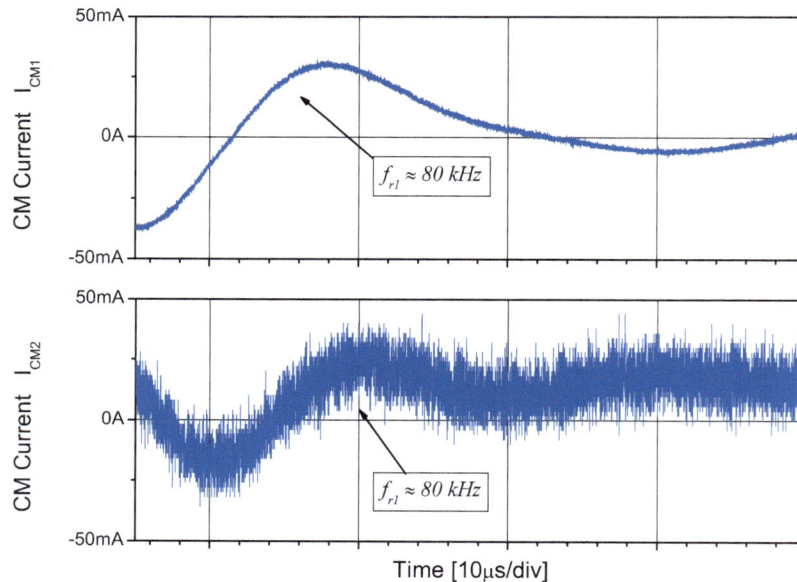

Figure 15: Comparison of the measured LF components of line side I_{CM1} and load side I_{CM2} CM currents of the investigated converter with AC motor fed by short cable.

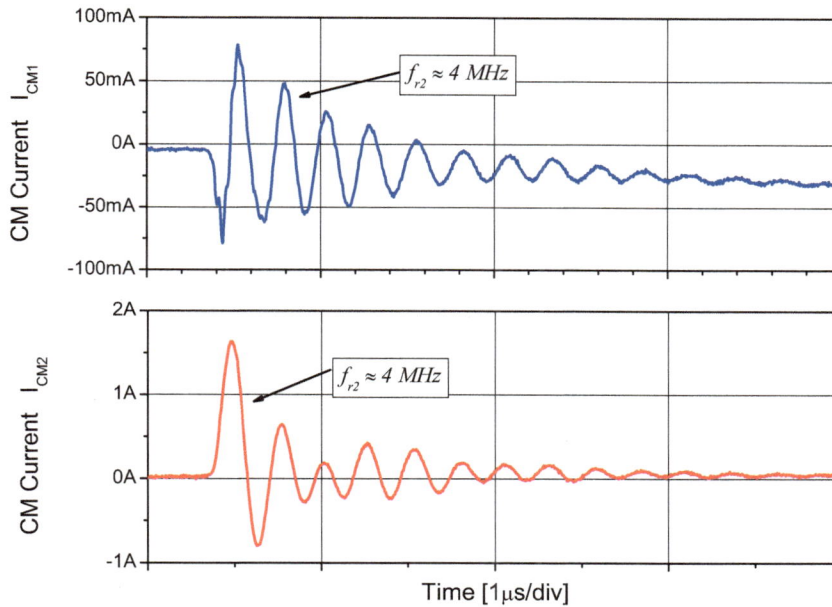

Figure 16: Comparison of the measured HF components of line side I_{CM1} and load side I_{CM2} CM currents of the investigated converter with AC motor fed by a short cable.

The effect of applying long feeding cable in the time domain is shown in Fig. **17** with reference to the case where the feeding cable was short. In this figure, selected single CM current pulses measured at the converters load side with possibly similar transistor switching conditions are presented.

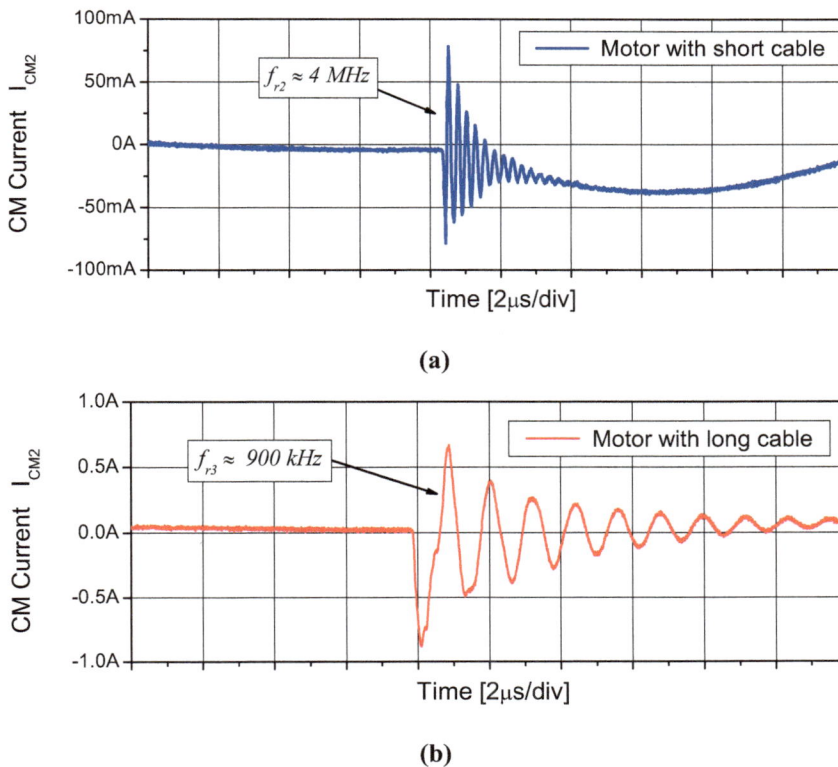

(a)

(b)

Figure 17: Comparison of the measured output CM currents for the AC motor fed by a short and a long cables.

Considerable differences of magnitudes and ringing frequencies of these CM currents can be clearly seen. Ringing frequency for the long cable case is approximately five times (~4 MHz against ~0.9 MHz) lower and tightly correlated with the resonance observed at the adequate impedance characteristic. A difference between the magnitudes of the CM currents is about ten times which shows a larger CM current for the long cable.

Analysis of the CM currents in frequency domain is more effective, because different CM current ringings, occurring at within modulation period are averaged and weighted in the detector of a spectrum analyzer during relatively long measuring time and are represented as averaged power spectrum density. A comparison of the measurement results in the frequency domain measured by an EMI receiver is presented in Fig. **18**, which verifies the input CM currents in correlation with the load impedance.

The measured CM current emission spectrum for the AC motor with a short cable exhibits a significant frequency region with elevated emission levels, located in the frequency range around *4 MHz*. This frequency range is closely correlated with the motor common mode impedance characteristic presented in Fig. **10**. For the AC motor with a long cable at least few particular frequencies ranges with evidently elevated emission levels are possible to determine. The most visible frequency ranges with elevated emission are approximately at *0.9 MHz, 2 MHz, 3 MHz, 6 MHz*. These frequencies are reliably correlated well with the adequate resonance frequencies observed at the CM impedance characteristic of the motor with the long cable presented in Fig. **10**.

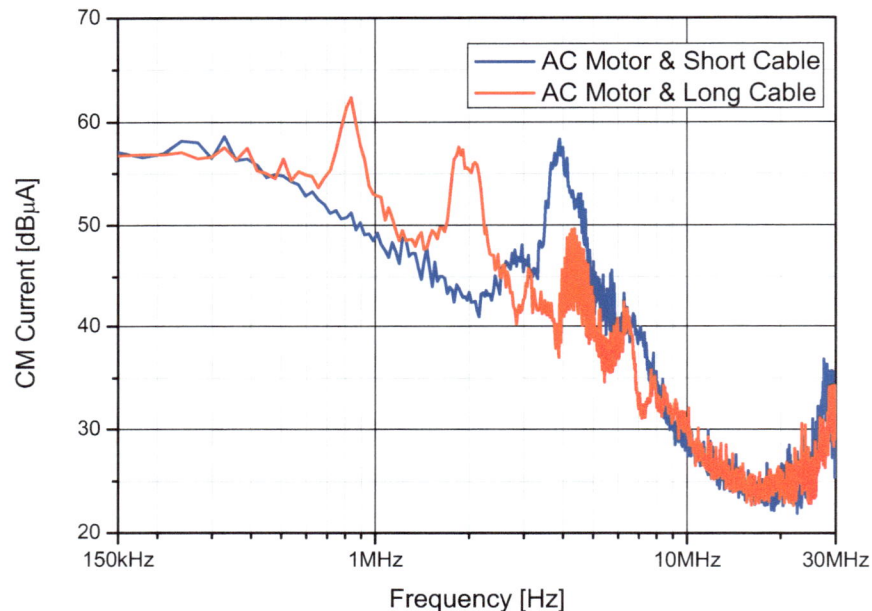

Figure 18: Comparison of the input CM currents emission spectra levels of the evaluated ASD for a short and a long motor feeding cables.

DESTRUCTIVE CONSEQUENCES OF PWM INVERTER FEEDING FOR AC MOTORS

Nowadays electric drives with induction motors fed by frequency converters are utilized very intensively in many applications due to their numerous advantages. Unfortunately, there are also some serious inconveniences of using frequency converters in AC drive applications, which need to be limited. Many of currently carried out research investigations are focused on converters design to reach high power and efficiency, low ratio of volume to power and to minimize the overall cost. These requirements force the utilization of higher magnitude of commutated DC voltage bus as well as higher modulation frequencies and more advanced control strategies, while maintaining the switching losses within the reasonable range. These trends lead unavoidably to high levels of steepness of converter output voltage changes *dv/dt*, which induce many adverse effects for the supplied AC motors such as harmful high frequency capacitive CM currents.

Voltage Effects of PWM Modulation for AC Motor Windings

AC motor windings fed by a frequency converter are exposed to steep voltage changes with a broadband spectral content which results in a high frequency voltage ringing. This phenomenon is associated with distributed parasitic capacitances of motor windings and a feeding cable. All the distributed parasitic capacitances of the windings are energized by the steep voltage pulses formed by the converter voltage oscillations. These oscillations appear due to voltage reflections phenomena taking place at the impedance mismatches inside the windings and at the most meaningful impedance mismatch at the winding terminals. An example of a motor voltage waveform with visible over voltage transients, which can reach to doubled the DC bus voltage value is presented in Fig. **19**.

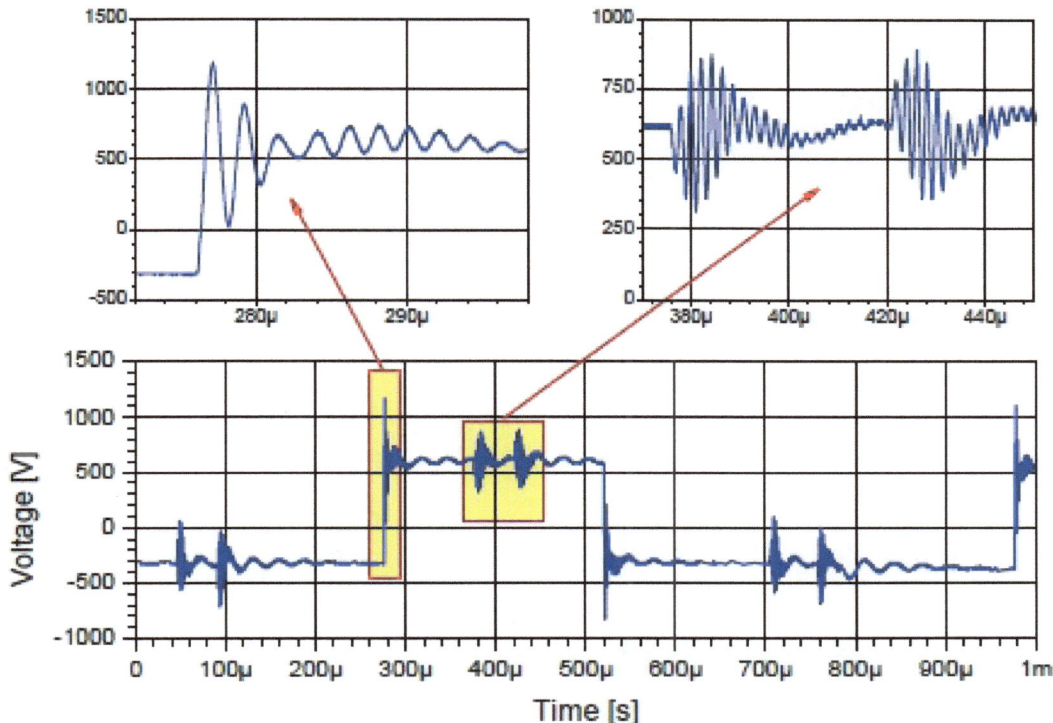

Figure 19: The exemplary waveform of a voltage ringing at AC motor terminals fed by converter.

Common Mode Currents in AC Motors Fed by Converter

Common mode currents flowing through an AC motor are directly related to the motor common mode voltage provided unintentionally by voltage source PWM converter and motor's common mode impedance. For high frequencies, the motor's CM impedance is hardly related to the winding's parasitic capacitances, especially the winding to ground capacitances. Depending on the analysed frequency range, equivalent parasitic winding to ground capacitance can differ noticeably because of its distributed character. Experimental investigations prove that with the increase of frequency from *9 kHz* up to *30 MHz* the value of winding equivalent parasitic capacitance can change few times [9]. This dependence complicates noticeably analysis and necessitates using distributed model.

The exemplary results of CM impedance measurement for a typical winding of an AC motor with different rated power are presented in Fig. **20**. Based on the CM impedance-frequency characteristics of the AC motor windings it can be noticed that equivalent capacitance changes noticeably at least few times within the frequency range of *9 kHz - 30 MHz*. The lowest CM impedance appears in relatively high frequency range above *1 MHz*.

The equivalent parasitic capacitance of the AC motor winding varies with frequency due to its distributed character. Therefore, detailed correlation of CM currents with CM voltage requires using distributed model presented in Fig. **6**.

Time and magnitude correlation of the CM current pulses with the CM voltage measured for a typical ASD is presented in Fig. **21**. The highest CM currents are generated during the transistor switching times when the voltage gradient *dv/dt* has maximum value.

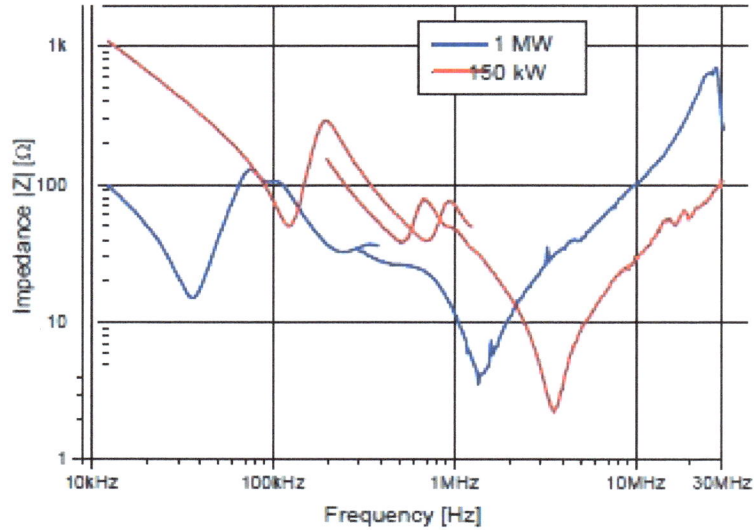

Figure 20: The common mode windings impedance of AC motor with different power.

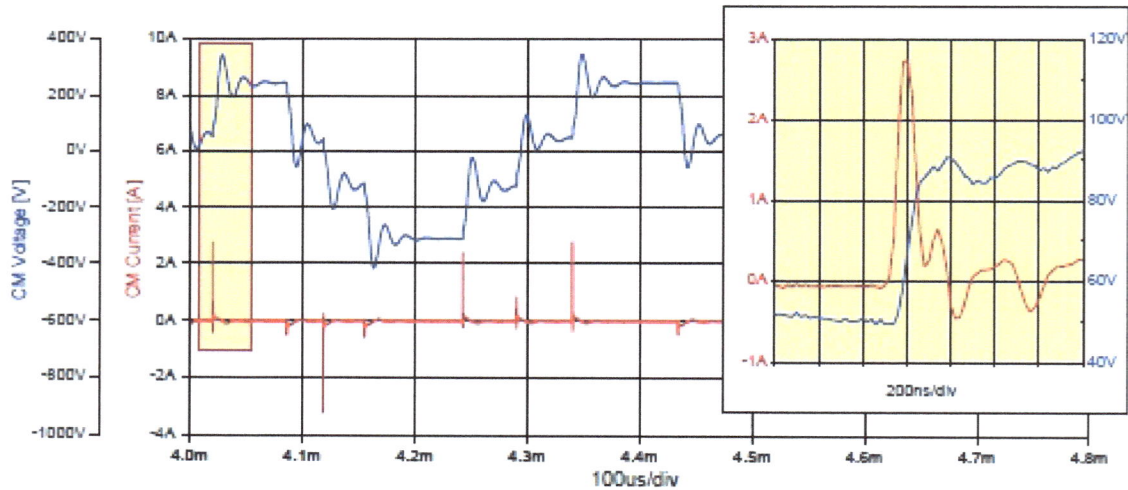

Figure 21: The correlation between the CM voltage and the CM current of an AC motor fed by a converter.

Shaft Voltages and Bearings Currents

The CM voltage at the motor terminals with a high level of *dv/dt* generates high frequency CM current components, which flow different ways between the motor windings, the grounded stator and the rotor through parasitic capacitances. The most important parasitic capacitances existing in a typical AC motor are presented in Fig. **22** and a simplified equivalent circuit model is shown in Fig. **23**.

According to the simplified circuit model (Fig. **23**), a part of the rotor to frame common mode current can also flow through the motor bearings. The bearing voltage $V_B(s)$ can be estimated by formula (9), on the basis of capacitive voltage divider build of C_{WR} and C_{RF} in accordance with Fig. **23**. If shaft voltage exceeds a breakdown voltage, there be a short circuit and it can discharge the energy stored in C_{RF} through the ball bearing and damage it.

$$V_B(s) = V_{CM}(s)\frac{C_{WR}}{C_{WR} + C_{RF} + C_{B1} + C_{B2}}$$

(9)

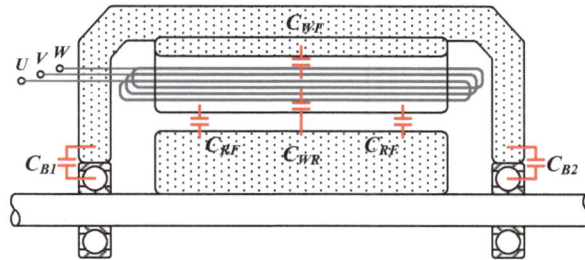

Figure 22: The parasitic capacitive couplings between AC motor windings: C_{WF} - winding to frame, C_{WR} - winding to rotor, C_{RF} - rotor to frame, C_{B1} and C_{B2}- bearing capacitances.

Example of measured bearing currents and correlation between the overall AC motor CM current and bearing current are presented in in Fig. **24**. According to the presented results the capacitive bearing current magnitude reach about 10 % of the total motor CM currents.

If the shaft voltage exceeds a critical bearing threshold voltage which breaks down the insulating grease thin film, the electric charge accumulated between the rotor and the frame is then unloaded through the bearing in the form of a discharging current (see Fig. **25**). Bearing current causes many problems such as fluting and pitting in the ball bearing and races structure. It affects the bearing lifetime and increase the maintained cost of the electric motors.

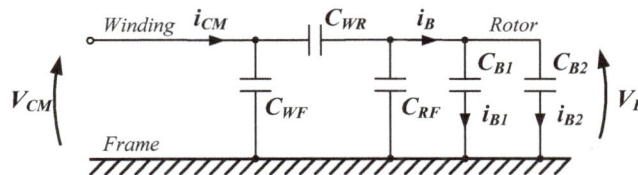

Figure 23: The simplified equivalent circuit model of capacitive couplings between AC motor windings, stator and rotor.

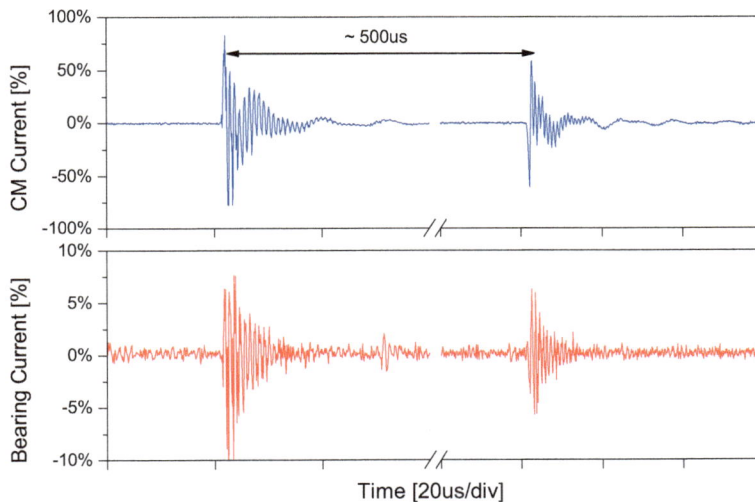

Figure 24: Relationship between the overall AC motor CM current and bearing current.

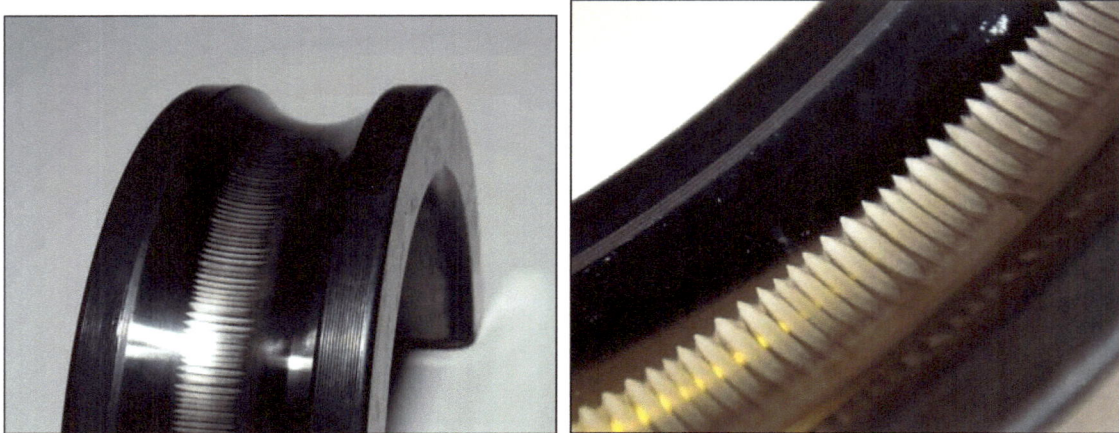

Figure 25: Examples of specific damages of an AC motor bearing's raceway by bearing currents.

CONCLUSIONS

The conducted EMI emission of ASD introduced into the power grid depends noticeably on the load parameters. A correlation between the line and the load side CM currents can be analysed more effectively by determining the converter load CM impedance - frequency characteristic. The AC motor windings with a feeding cable, and its CM impedance-frequency characteristic have fundamental influence on the CM current emission levels, in particular in the upper part of CISPR B conducted EMI frequency band (150 kHz - 30 MHz). Typical AC motor windings in the range of several hundreds of kW have one or more self resonance frequencies in the conducted EMI frequency range.

An important part of the load is the motor feeding cable which introduces extra resonances in the converter load impedance characteristic due to reflection phenomena at motor terminals where the impedance mismatch is dominant. The reflections are a result of impedance mismatch between the feeding cable and the AC motor windings. With the increase of motor feeding cable length the minimum frequency of the observed resonances decreases and the number of appearing resonances increases. In addition the motor cable slightly reduces the maximum emission levels related with the motor self resonance and introduces extra emission increase in the lower frequency range. The proposed wideband circuit model of the converter's load can be successfully used for optimization of AC motor feeding cable broadband interaction with motor's windings for different purposes, especially for determining of CM currents EMI emission spectra.

REFERENCES

[1]   Ogasawara S, Akagi H. Analysis and Reduction of EMI Conducted by a PWM Inverted-Fed AC Motor Drive System Having Long Power Cables. Proc. of IEEE 31st Annual Power Electronics Specialists Conference (PESC'00), Vol. 2, 2000; pp. 928-933.

[2]   Adabi J, Zare F, Ledwich G, Ghosh A. Leakage Current and Common Mode Voltage Issues in Modern AC Drive Systems, AUPEC, Perth, Australia, Dec 2007.

[3]   Łuszcz J, Iwan K. AC motor transients and EMI emission analysis in the ASD by parasitic resonance effects identification, European Conference on Power Electronics and Applications, EPE 2007.

[4]   Zare F. Modelling of Electric Motors for Electromagnetic Compatibility Analysis, AUPEC 2006, Melbourne, Australia, Nov 2006.

[5]   Said, Al-Haddad K. A new approach to analyze the overvoltages due to the cable lengths and EMI on adjustable speed drive motors, Power Electronics Specialists Conference, 2004. PESC 04. 2004 IEEE 35th Annual, Volume 5, 20-25 June 2004; 5: 3964 - 3970.

[6]   Weidinger T. Elimination of increased excitation of common-mode oscillations in electrical drive systems with active front end and long motor cables, Power Electronics and Motion Control Conference, 2008. EPE-PEMC 2008.

[7]   Zare F. High frequency model of an electric motor based on measurement results, Aust J Elect Electron Eng (AJEEE) 2008; 4 (1): 17-24.

[8]   Arnedo L, Venkatesan K. High frequency modelling of induction motor drives for EMI and overvoltage mitigation studies. Electric Mach Drives Conf (IEMDC'03) 2003; 1(1-4): 468-474.

[9]   Łuszcz J, Mosoń I. AC Motor Windings Circuit Model for Common Mode EMI Currents Analysis, The 5th International Conference CPE 2007, Compatibility in Power Electronics, May 29 - June 1, 2007, Gdansk, Poland.

[10]   Moreira AF, Lipo TA, Venkataramanan G, Bernet S. High-frequency modelling for cable and induction motor over voltage studies in long cable drives. IEEE Trans Indus Appl 2002; 38(5): 1297-1306.

[11]   Hanigovszki N, Poulsen J, Spiazzi G, Blaabjerg F. An EMC evaluation of the use of unshielded motor cables in AC adjustable speed drive applications, Power Electronics Specialists Conference, 2004. PESC 2004. Volume 1, 20-25 June 2004 Page(s): 75-81 Vol. 1.

[12]   Łuszcz J. Motor Cable as an Origin of Supplementary Conducted EMI Emission of ASD, 13th European Conference on Power Electronics and Applications (EPE 2009), 8 - 10 September 2009, Barcelona, Spain.

[13]   Łuszcz J. Motor Cable Influence on the Converter Fed AC Motor Drive Conducted EMI Emission. The 5th International Conference CPE 2009, Compatibility and Power Electronics, May 20 - 22, 2009, Badajoz, Spain.

[14]   Muetze, Binder A. Calculation of Circulating Bearing Currents in Machines of Inverter-Based Drive Systems. IEEE Trans Indust Electron April 2007; 54(2): 932 - 938.

[15]   Akagi H, Tamura S. A passive EMI filter for elimination both bearing current and ground leakage current from an inverter-driven motor. IEEE Trans Power Electronics Sep. 2006; 21(5): 1459-1469.

[16]   Adabi J, Zare F, Ghosh A. End-winding Effect on Shaft Voltage in AC Generators, 13th European Conference on Power Electronics and Applications (EPE 2009), 8-10 September 2009, Barcelona, Spain.

[17]   Kempski, Strzelecki R, Smolenski R, Fedyczak Z. Bearing current path and pulse rate in PWM-inverter-fed induction motor, PESC. 2001 IEEE, Volume: 4, 17-21 June 2001 Pages: 2025 - 2030 vol. 4.

[18]   Costabile G, Vivo B, Egiziano L, Tucci V, Vitelli M, Beneduce L, Iovieno S, Masucci A. An accurate evaluation of electric discharge machining bearings currents in inverter-driven induction motors, European Conference on Power Electronics and Applications (EPE 2007) 2-5 Sept. 2007.

[19]   Paul CR Introduction to Electromagnetic Compatibility. New York: John Wiley & Sons, 1992.

CHAPTER 5

Lightning Surges on Wind Power Systems

R.B. Rodrigues[1], V.M.F. Mendes[1] and J.P.S. Catalão[2,*]

[1]Department of Electrical Engineering and Automation, Instituto Superior de Engenharia de Lisboa, R. Conselheiro Emídio Navarro, 1950-062 Lisbon, Portugal and [2]Department of Electromechanical Engineering, University of Beira Interior, R. Fonte do Lameiro, 6201-001 Covilha, Portugal

Abstract: As wind power generation undergoes rapid growth, lightning damages involving wind power systems have come to be regarded as a serious problem. This chapter gives an introduction to lightning phenomena, lightning location systems and important parameters regarding lightning protection. The wind turbine is described and the main characteristics of its components, like the tower, the generator, the blades, and the electrical and electronic equipment, are highlighted. This chapter also introduces fundamentals of risk analysis method based on international standards, and describes how the rolling sphere method can be used to identify the vulnerable points on a structure. Computer tools and simulations using the LPS 2008 computer program are presented and discussed. Finally, wind turbine issues are discussed.

Keywords: Transmission system, Blades, Control and safety systems, Grid connection, Generator, Lightning phenomena, Lightning location systems, Lightning surge parameters, Protection issues in wind turbine, Risk analysis, Wind turbine protection,

INTRODUCTION TO LIGHTNING

The need to control climate changes and the increase in fossil-fuel costs stimulate the ever-growing use of renewable energies worldwide. Concerning renewable energies, wind power is a priority in energy strategy for many countries.

As wind power generation undergoes rapid growth, lightning incidents involving wind power systems have come to be regarded as a serious problem [1]. Lightning protection of wind turbines presents problems that are not normally seen with other structures. These problems are a result of the following issues [2]:

- Wind turbines are tall structures of up to more than 150 m in height;

- Wind turbines are frequently placed at exposed locations to lightning strokes;

- The blades and nacelle are rotating;

- The most exposed wind turbine components such as the blades and the nacelle are often made of composite materials incapable of sustaining direct lightning stroke or of conducting lightning current;

- The lightning current has to be conducted through the wind turbine structure to the ground, whereby significant parts of the lightning current will pass through or near to practically all wind turbine components;

- Wind turbines in wind farms are electrically interconnected and often placed at locations with poor earthing conditions.

Modern wind turbines are characterized not only by greater heights but also by the presence of ever-increasing control and power electronics. Consequently, the design of the lightning protection of modern wind turbines could be a challenging problem [3]. The future development of wind power generation and the construction of more wind farms will necessitate intensified discussion of lightning protection and the insulation design of such facilities [4].

This book chapter is organized in three parts: Part 1 – Introduction to lightning. In Part 1 a review concerning lightning phenomena, lightning location systems and lightning surge parameters are presented. Part 2 – Wind turbine generator system. In Part 2 the wind turbine generation system is described. Part 3 – Protection issues in wind turbine. Part 3

*Address correspondence to J.P.S. Catalão: Department of Electromechanical Engineering, University of Beira Interior, R. Fonte do Lameiro, 6201-001 Covilha, Portugal; Tel: 00351275329972; E-mail: catalao@ubi.pt

underlines the importance of risk analysis and the use of a practical and well established method to determine vulnerable points on a structure. The advantages of software tools on risk analysis, and three-dimensional (3D) simulation of the Rolling Sphere Method, are also discussed. Finally, the necessary protection measures and conclusions are duly drawn.

Lightning Phenomena

Lightning has been faced with respect and a symbol of power for many civilizations. In the ancient Egypt, Typhon was the God responsible to throw the rays from the sky. Ancient books of Vedas in India describe Indra as the God of the sky, the rays, the rain and the storms. Indra is represented inside a car which carries the rays. Sumerian people represented the God Zarpenik, in 2500 B.C., riding a horse with rays on both hands. The Greeks believe that the ray was a weapon used by Zeus and his family. In Greek mythology the ray was manufactured by Minerva, God of wisdom. Lightning was seen by Greeks and Romans like a divine phenomena and the place touched by them was sacred soil, which is why many temples were built in places stroked by lightning. Séneca wrote a book about lightning, probably the first one, where we can read: "Júpiter throw the rays against columns, trees and sometimes even against their own statues".

Lightning is in fact a tremendous phenomenon of nature, as beautiful as dangerous. Fig. **1** shows Lisbon under a storm in 2009. The lightning phenomenon is associated to the formation of clouds. When a humid air mass is heated and raises on the air the cloud is formed. The hot and humid mass of air can be raised by several mechanisms, as shown in Fig. **2**.

Figure 1: Lightning over Lisbon in 2009.

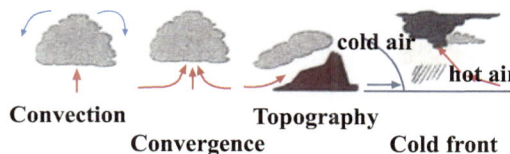

Figure 2: Mechanisms for raising a humid air mass.

While the humid air raises slowly, the atmospheric pressure drops, the mass of air enlarges and the temperature drops. If this process continues being fed by humid air and the mass of air continues rising, the temperature can drop to dewpoint. When dewpoint occurs the water vapour suffers condensation around hygroscopic nucleus and the cloud is formed. Hygroscopic nucleus can be particles of dust, smoke or other microscopic substances. When the temperature reaches the dewpoint, the water vapour move from gas to liquid and releases the latent heat. This release of heat slows the temperature drop inside the mass of air and contributes to continue lifting up the cloud. Climate conditions can produce different types of clouds (Fig. **3**).

Figure 3: Common types of clouds in the troposphere.

Clouds which can produce lightning are called cumulonimbus. The cumulonimbus has the basis typically around 700 and 1500 m of altitude and can grow until 24 or 35 km, but most of the times it has 9 to 12 km of height. Inside, the cumulonimbus has drops of water on the basis and crystals of ice and snow on the top, where temperature is around -50 to -70 °C. Strong winds inside the cloud can cause enormous turbulence. Winds are caused by hot air going up and cold air going down. When the top of the cloud reaches the troposphere, strong winds and the change on temperature forces the top of the cloud horizontally, giving the well known aspect of bighorn.

The planet Earth is electrically charged and its behaviour can be compared to a spherical capacitor. The solid part of the Earth has a negative charge of 1×10^6 C while the atmosphere has identical charge but positive [5]. Observations indicate that at least 1000 to 2000 thunderstorms occur every minute on our planet [6]. Lightning can be compared to an electric generator which injects current into the ground to restore the electric stability disturbed by the process of formation of the cumulonimbus.

Lightning has been classified in: Intra-Cloud (IC), Cloud-Ground (CG), Inter-Cloud, Cloud-to-Air and lightning discharges in the upper atmosphere [7]. The IC lightning is the most frequent one and occurs between the negative region on the basis of the cloud, and the positive region on the top. Most of the CG strokes start with an IC discharge. The stepped leader appears 1/10 of second after the IC. Fig. **4** shows the different stages of a downward negative CG discharge.

There is still no agreement on how the electric charge is formed or distributed inside the cloud, neither about the preliminary breakdown phase. However, when the electric field reaches 10 kV/m the process takes place. During the stepped leader phase a channel of plasma propagates to the ground at 2×10^5 ms^{-1}.

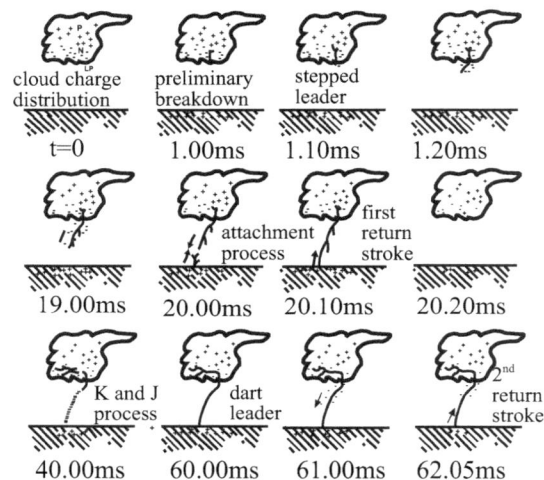

Figure 4: Stages of a downward negative CG discharge [7].

The stepped leader propagates by jumps with the duration of 1 µs and pauses of 20 to 50 µs. As the stepped leader approaches to the ground, conditions are created to initiate upward streamers, usually starting in tall structures like buildings or wind turbines. One of the multiple streamers will succeed and attach to the stepped leader. At this time the first return stroke occurs. The first return stroke is a massive and positive current propagating upward at half the speed of light which uncharges the ionized channel.

Lightning photographed with one still camera and another moving is shown in Fig. **5**. If conditions of charge are favourable after 40 to 80 ms a dart leader can re-establish the channel and a subsequent return stroke can happen.

After the first return stroke some subsequent return strokes can happen while the balance of electric charge is not re-established. During the processes of ionization and disruption strong emissions of radiation in the range of very high frequency (VHF) are observed. However, during the return strokes phases most of radiation is at low and very low frequency (LF, VLF). Fig. **6** was presented by Malan, showing the relationship between frequency and lightning detection method [8].

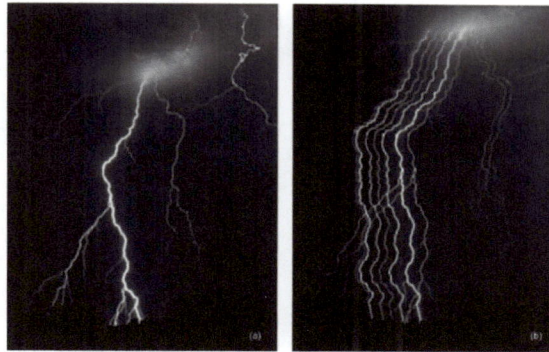

Figure 5: Lightning photographed: **a)** still camera (on the left); **b)** camera in movement (on the right) [7].

During CG strikes, there are only a few impulses of considerable magnitude in the range of VLF and LF signals, which are a function of the channel length and the peak current [9]. On the other hand, IC strikes produce hundreds of small impulses in LF, but in the range of VHF the magnitudes of the signals emitted are comparable to the VLF and LF signals of CG strikes.

Lightning Location Systems

Lightning Location Systems (LLS) are of great importance for the design of more effective lightning protection systems. Actually LLS are used to characterize lightning activity in particular regions, to trace and predict the direction of thunderstorms and to measure their intensity. This allows the construction of Ground Flash Density (GFD) maps to be much more accurate than Thunderstorm Days (Td) maps.

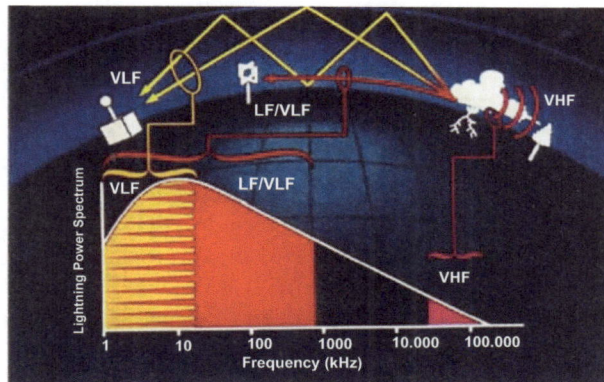

Figure 6: Relationship between frequency and lightning detection method [8].

Accurate GFD maps are essential for accurate risk analysis and consequently to choose the most adequate protection measures to a particular situation.

The information of LLS is also used by meteorologists as an additional source of information to improve the lightning predictions based on data of meteorological radars and satellite images.

Several methods and systems have been developed to detect and locate the CG strikes. The Direction Finders (DF) system operates in the range of 1 to 500 kHz (VLF and LF). It was designed to intercept signals from CG return strokes and each detector is able to cover 500 km around it [10].

The electric field is detected to determine polarity of lightning and the peak current is estimated as a function of the magnitude of magnetic field. The location of the point of impact can be achieved by triangulation, as shown in Fig. 7.

The DF system has difficulties when the lightning strike occurs in line between two detectors and these two are the only ones to detect it.

Lewis *et al.* [11] developed a system based on the Time-Of-Arrival (TOA). This system requires precision clocks on each detector, precisely synchronized, to measure the instant of detection.

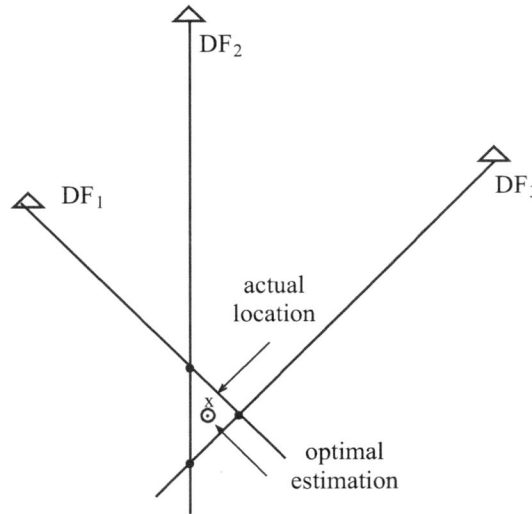

Figure 7: Optimal scenario for DF system [10].

Being known the speed of electromagnetic waves in the air, a difference in the arrival times at two stations defines a hyperbola, and multiple stations provide multiple hyperbolas whose intersections define the source location. This technique is illustrated in Fig. **8**. Under some geometrical conditions, curves produced from only three detectors will result in two intersections, leading to an ambiguous location. This problem is avoided if four lightning strike detectors are used.

TOA systems can provide accurate locations at long ranges [12], and if the antennas are properly sited, the systematic errors are minimal. Casper and Bent [13] have developed a wideband TOA receiver, the Lightning Position and Tracking System (LPATS) that is suitable for locating lightning sources at medium and long ranges using the hyperbolic method.

More recently the Global Atmospherics Inc. has developed the so-called Improved Accuracy Using Combined Technology (IMPACT) system which combines the DF and TOA techniques. This methodology assures redundant information, allowing a more precise detection even in adverse geometric situations. Fig. **9** shows an example to detect the location of the impact point with three IMPACT detectors and two LPATS TOA.

In Fig. **9** the direction measurements are shown as straight-line vectors, and range circles centred on each detector represent the TOA measurements.

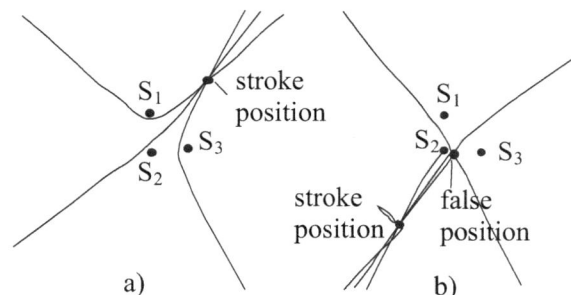

Figure 8: a) Optimal scenario for TOA system (on the left); **b**) Ambiguous scenario for TOA system (on the right) [11].

Direction finding based on VHF interferometry, and Lightning Detection and Ranging (LDAR) system that is capable of providing three-dimensional locations of more than a thousand RF pulses within each lightning flash, are the next generations of the lightning location systems that are being developed.

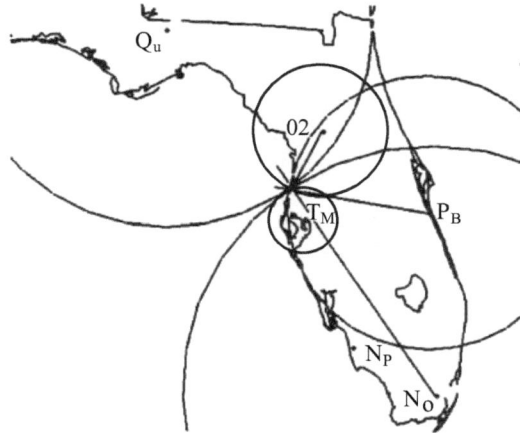

Figure 9: Example of location with IMPACT system [11].

Lightning Surge Parameters

Lightning can be regarded as a current source as far as the direct effects are concerned and it appears more as an electromagnetic source for indirect effects.

In flat territory and to lower structures, mostly downward flashes occur, whereas for exposed and/or higher structures like wind turbines, upward flashes become dominant. With the effective height, striking probability increases and the physical conditions change.

A lightning current consists of one or more different strokes:

- Short strokes with duration less than 2 ms (Fig. **10**, where: O_1 = virtual origin; I = peak current; T_1 = front time; T_2 = time to half value).

- Long strokes with duration longer than 2 ms (Fig. **11**, where T_{long} = duration time; Q_{long} = long stroke charge).

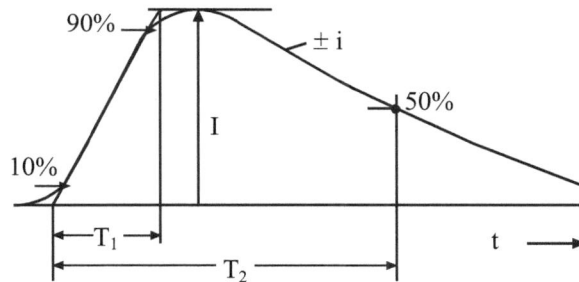

Figure 10: Definitions of short stroke parameters (Typically $T_2 < 2$ ms) [14].

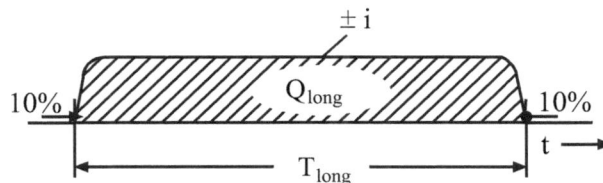

Figure 11: Definitions of long stroke parameters (Typically 2 ms $< T_{long} < 1$ s) [14].

Return strokes can combine in different ways, as shown in Fig. **12**.

The IEC 62305 series are a set of standards that provide the general principles to be followed in the protection against lightning of:

       –   Structures including their installations and contents as well as persons;

       –   Services entering a structure.

The following cases are outside the object of this standard: railway systems; vehicles, ships, aircraft, offshore installations; underground high pressure pipelines. Usually these systems are under special regulations made by various specific authorities.

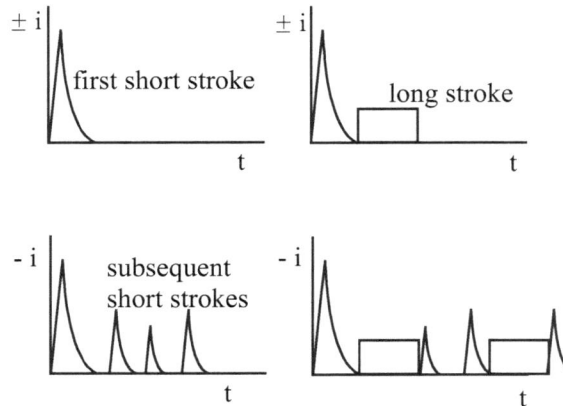

Figure 12: Possible components of downward flashes (typical in flat territory and to lower structures) [14].

When a lightning strikes a structure, the current wave propagates through the structure to the ground and generates a corresponding voltage surge which will be a function of the current wave and the characteristic impulsive impedance of the structure.

If a lightning hits an electrical power line, it causes two voltage surges which will propagate in both directions at a velocity dependent on the line impedance, as shown in Fig. **13** (where U = peak voltage; Z_0 = impulsive impedance of the line) and in the following equation:

$$v = \frac{1}{\sqrt{\mu\varepsilon}} = \frac{1}{\sqrt{LC}} < c = \frac{1}{\sqrt{\mu_0\varepsilon_0}} \tag{1}$$

where: v = velocity of voltage wave; μ_0, μ_r and μ = magnetic permeability of vacuum, relative and absolute magnetic permeability of the line; ε_0, ε_r and ε = electric permittivity of vacuum, relative and absolute electric permittivity of the line; L and C = inductance and capacitance of the line, respectively; c = speed of the light.

When the voltage wave arrives to the first pole the voltage can exceed the breakdown voltage supported by the insulators and a back flashover happens.

The current wave shapes of the first short stroke 10/350 μs and the subsequent short strokes 0,25/100 μs may be defined [14] as:

$$i = \frac{I}{k} \cdot \frac{(t/\tau_1)^{10}}{1+(t/\tau_1)^{10}} \cdot \exp(-t/\tau_2) \tag{2}$$

where: I = peak current; k = correction factor for the peak current; t = time; τ_1 = front time constant; τ_2 = tail time constant. Values of lightning parameters are usually obtained from measurement taken on tall objects.

Statistical distribution of estimated lightning current peak values that does not consider the effect of tall objects is also available from local LLS.

$$U = \frac{I}{2} \cdot Z_0$$

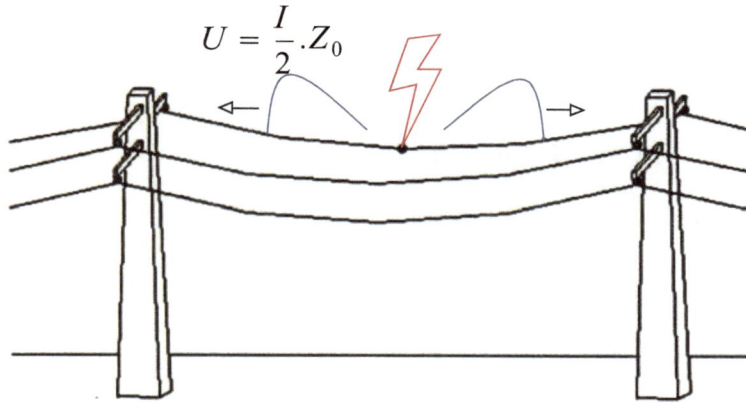

Figure 13: Propagation of voltage surges along a power line.

The parameters of the current wave shapes of the first and the subsequent short strokes with different Lightning Protection Levels (LPL), are given in Table 1 [14]. All values given in this standard relate to both downward and upward flashes.

WIND TURBINE GENERATION SYSTEM

The Blades

The three bladed rotor is the most important and visible part of the wind turbine. The wind energy is transformed into mechanical energy, which turns the main shaft of the wind turbine. The front and rear sides of a wind turbine rotor blade have a shape roughly similar to a long rectangle, with the edges bounded by the leading edge, the trailing edge, the blade tip and the blade root. The blade root is bolted to the hub. The cross section has a streamlined asymmetrical shape, with the flattest side facing the oncoming air flow or wind.

Table 1: Parameters for equation (2)

Parameters	First short stroke			Subsequent short stroke		
	LPL			LPL		
	I	II	III - IV	I	II	III -IV
I (kA)	200	150	100	50	37.5	25
k	0.93	0.93	0.93	0.993	0.993	0.993
τ_1 (μs)	19.0	19.0	19.0	0.454	0.454	0.454
τ_2 (μs)	485	485	485	143	143	143

This shape is called the blade's aerodynamic profile. Modern blades are manufactured in a vacuum process and a sandwich method. Layers of fibreglass are placed in the mold with resin. The vacuum process assures that no air bubbles remain inside the blade. Fig. **14** shows images and parts of a blade.

The world's largest turbine, produced by Enercon in 2008, has a blade diameter of 126 m and a rated power of 6 MW, but will most likely produce more than 7 MW.

The Transmission System

The transmission system is the link between the wind turbine blades and the generator. The transmission system is constituted by: the hub, the main shaft, the main bearings, the clamping unit, the gearbox and the coupling unit. Fig **15** shows how these components are placed together.

More recently a transmission system avoiding the main shaft and the gearbox has been used, reducing considerably the size and the number of parts.

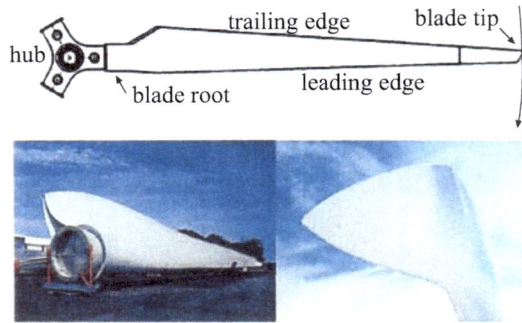

Figure 14: Parts and images of a blade (Source: Enercon).

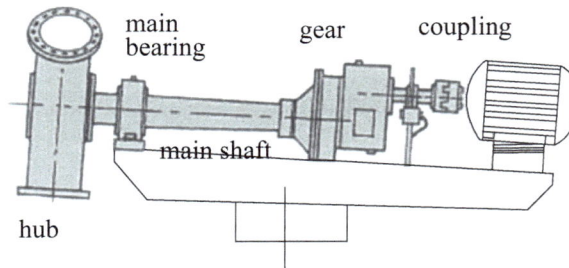

Figure 15: The link between the wind turbine blades and the generator.

Fewer rotating components reduce mechanical stress and at the same time increase the technical service life of the equipment. The hub is connected directly to the annular generator and only two bearings are used. Fig **16** shows a representation of this system.

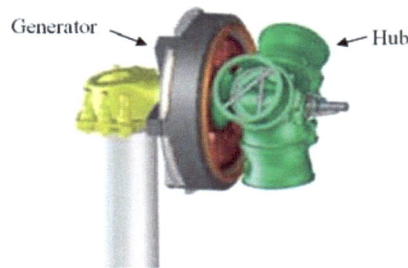

Figure 16: Transmission system without main shaft and gearbox (Source: Enercon).

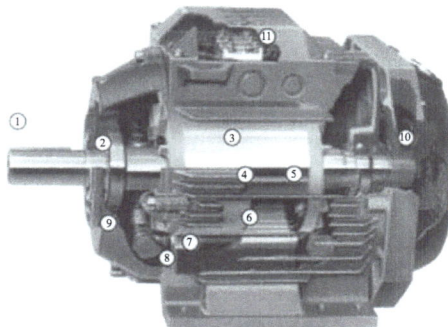

1. Generator shaft	7. Coil
2. Rolling bearings	8. Stator plates
3. Rotor	9. Coil heads
4. Rotor aluminium bar	10. Ventilator
5. Rotor aluminium ring	11. Connection box
6. Stator	

Figure 17: Components of an induction generator.

The Generator

The generator is the unit of the wind turbine that transforms mechanical energy into electrical energy. The asynchronous generator is a very common type of generator used in wind turbines. It is often referred as an induction generator. Fig **17** shows a representation of the induction generator.

The annular generator (Fig. **18**) is a low-speed synchronous generator with no direct grid coupling.

Output voltage and frequency vary with the speed and are converted for output to the grid *via* a DC link and a converter achieving high speed variability.

The Control and Safety Systems

Control and safety systems of a wind turbine comprise of many different components, mostly based on electrical and electronic equipment.

Combined together they are part of a more comprehensive system, insuring that the wind turbine is operated satisfactory and preventing possible dangerous situations.

Fig. **19** shows the simplified block diagram of grid connected wind turbine with the control and the measuring systems.

Figure 18: Annular generator under manufacturing (Source: Enercon).

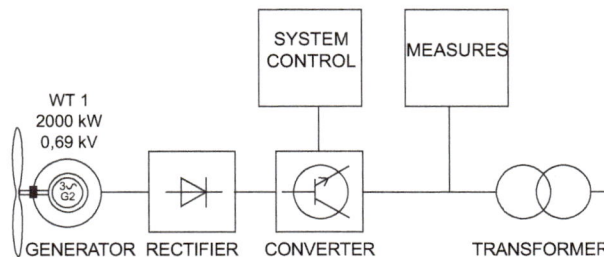

Figure 19: A grid connected wind turbine.

Ensuring proper power feed from wind turbines into the grid requires grid connection monitoring. Grid parameters such as voltage, current and frequency are measured on the low-voltage side between the converter and the power transformer. The measured values are transmitted to the control system, enabling the turbine to react immediately to changes.

Some wind turbines are equipped with a storm control system, which enables reduced turbine operation in the event of extremely high wind speeds. This is achieved by turning the rotor blades slightly out of the wind. When the defined limit values for system or grid protection are exceeded, the wind turbine is securely shut down and the service teams are informed. As soon as conditions return within the permissible tolerance range, the wind turbine is automatically started up again at full power. This prevents frequent shutdowns and the resulting yield losses.

The Tower and Foundation

The foundation transmits the wind turbine's dead load and wind load into the ground. Circular foundations have the advantage that forces are equal in all directions. Depending on the site, the ground can only absorb a certain amount of compressive strain, so the foundation surfaces are adapted accordingly.

Fig. **20** shows the construction of a circular foundation for a tall steel tower.

The tower is the physical support for all components of a wind turbine like blades, transmission system, generator, and control and safety systems.

Two types of towers are usually installed: concrete armed towers and steel towers. Tubular steel towers are usually manufactured in several individual tower sections connected using stress reducing L-flanges. In steel towers, full corrosion protection quality must be of special concern for manufacturers.

Figure 20: Circular foundation for tall steel towers (Source: Enercon).

However, steel tower avoids the complicated and costly work on site necessary for a concrete armed tower. When a lightning hits a tower, the tower will be the most probable way for lightning current reaching the ground. A steel tower is itself a natural lightning rod and its considerable surface disperses the current, reducing the risk of damage.

The Grid Connection

Inside each tower a Low Voltage (LV) / High Voltage (HV) substation adapts the generator voltage output to the grid. Fig. **21** shows a typical electrical scheme for a wind turbine LV/HV substation.

Wind turbines are usually placed on a wind farm with a certain distance between them. This distance is evaluated taking into consideration the height and diameter of each wind turbine, orography and intensity of winds in the region.

Wind turbines are grouped and connected by a buried HV cable. Finally, all cables of the wind farm enter into a 20 kV/60 kV substation which delivers the electrical energy to the grid.

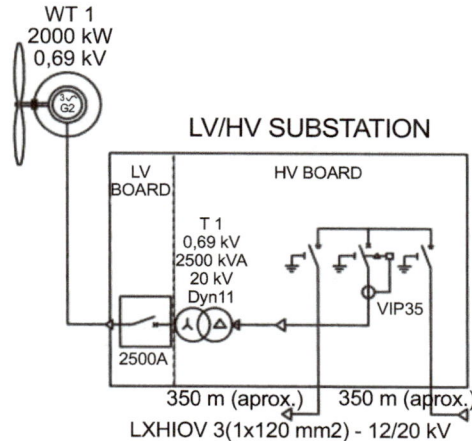

Figure 21: Typical scheme for a wind turbine LV/HV substation (Source: Siemens).

PROTECTION ISSUES IN WIND TURBINE

Damage events are registered in the databases as turbine faults caused directly or indirectly by lightning. A summary of these faults is shown in Table **2** for Germany, Denmark and Sweden.

The lightning caused faults vary from 3,9 to 8 events per 100 turbine years. Restated, in Northern Europe it is expected that 4 to 8 of every 100 turbines would be damaged by lightning in a given year [2].

When lightning hits a wind turbine without the proper protection, damages are often severe. Fig. **22** shows a wind turbine in Portugal damaged by lightning.

The economic benefits, the selection of adequate protection devices and other protection measures should be determined in terms of risk management.

Table 2: Lightning damage frequency [2]

Country	Period	Turbines in database	Capacity in MW	Turbine years	Lightning faults	Faults per 100 turbine year
Germany	1991-1998	1498	352	9204	738	8.0
Denmark	1990-1998	2839	698	22000	851	3.9
Sweden	1992-1998	428	178	1487	86	5.8

Figure 22: Wind turbine in Portugal damaged by lightning.

Risk analysis procedures have been reported in the literature [15], as to whether lightning protection is needed or not. However, there is currently no international standard specifically written for the lightning protection of wind turbines.

The closest document of the International Electrotechnical Commission (IEC), IEC TR 61400-24 [2], is purely informative and is not documented as an international standard, but it should be regarded as a significant breakthrough.

In January 2006, a new four-part standard document has appeared, IEC 62305-1 to 62305-4 [14–17], providing the general principles of protection against lightning. However, the standards IEC 62305 are not specifically developed for the lightning protection of wind power plants, since no account is taken for the peculiarities of wind turbines.

In fact, wind turbines with tall structures are often placed in very exposed regions to lightning. Additionally, they are composed by conductive parts in relative movement, what complicates the application of standard lightning protection system principles [18].

The risk management method reported in IEC 62305-2 [15] entails a significant number of input data and mathematical processing worthy of being supported by a computer program for a friendly practical application.

The identification of the most vulnerable points on a given structure to be struck by lightning is an important issue on the design of a reliable Lightning Protection System (LPS). The lightning strike model considered in series IEC 62305 is the rolling sphere method (RSM).

Risk Analysis

The method purposed in IEC 62305-2 [15] for risk assessment on a structure or on a incoming service like LV power line, telephone line or cable TV, due to lightning flashes to ground is based on a comparison of evaluated risk values.

Once an upper tolerable limit for the risk has been ascertained, this method allows the selection of appropriate protection measures in order to reduce the risk at or below the tolerable limit.

The following should be intended as a brief explanation of this method.

Lightning flashes influencing a structure or a service are considered sources of damage and are divided into:

- – S1 - flashes striking the structure;
- – S2 - flashes striking near the structure and/or near the connected services (power, telecom lines, other services);
- – S3 - flashes striking the service;
- – S4 - flashes striking near the service or direct to a structure connected to the service.

Lightning flashes can be hazardous to life, to structures and to services, and the following damages are considered:

- – D1 - injuries of living beings in or close to the structure;
- – D2 - damages to the structure and to its contents;
- – D3 - failures of associated electrical and electronic systems.

The sources of damages S1 and S3 may cause one of the damages: D1 - D3. The sources of damages S2 and S4 may cause D3.

Moreover, failures caused by lightning overvoltages in power supply lines may also generate switching overvoltages in the users´ installations like residential, commercial and industrial areas.

Each type of damage, alone or in combination with others, may produce a different consequent loss in the object which needs to be protected.

The type of loss that may appear depends on the characteristics of the structure itself and its content.

The following types of losses are taken into account:

- L1 - Loss of human life;
- L2 - Loss of service to the public;
- L3 - Loss of cultural heritage;
- L4 - Loss of economic value (structure and its content, service and loss of activity).

Protection measures may be required to reduce the loss due to lightning. Whether they are needed, and to what extent, should be properly determined by risk assessment.

The risk, defined in [15] as the average annual loss on a structure or on a service due to lightning flashes to ground, depends on:

- The annual number of lightning flashes influencing the structure and/or the service;
- The probability of damage by one of the influencing lightning flashes;
- The average amount of the consequent loss.

The number of lightning flashes influencing the structure or the service depends on their: dimensions and characteristics; environmental characteristics; as well as the lightning ground flash density in the region.

The probability of lightning damages depends on the structure, the service and the lightning current characteristics, as well as on the kind and efficiency of applied protection measures.

The annual average amount of the consequential loss depends on the extent of damages and the consequential effects, which may occur as a result of a lightning flash.

The effect of protection measures results from the features of each protection measure and allows for a reduction of the damage probability and the amount of consequential loss.

The probability of damage is statistically evaluated by (3):

$$F = (1 - e^{-\lambda \cdot t}) \tag{3}$$

Adapting (3) to the lightning phenomena, the mathematical expression to assess the risk value is given by (4):

$$R = (1 - e^{-NPt})L \tag{4}$$

where N is the number of lightning flashes per year influencing the structure/service; P is the probability of damage by one of the influencing lightning flashes; L is the consequent mean amount of loss; and t is the period under evaluation, which is usually considered one year.

Applying the MacLaurin series to (4) we find the general expression considered in [15] to evaluate the risk factors:

$$R = (1 - (e^{-N \cdot P \cdot 0}) + \left(\frac{-N \cdot P \cdot e^{-N \cdot P \cdot 0}}{1!} \right) \cdot t + \left(\frac{(-N \cdot P) \cdot (-N \cdot P) \cdot e^{-N \cdot P \cdot 0}}{2!} \right) \cdot t^2 + ...)L$$

$$= (1 - 1 - N \cdot P \cdot t + \left(\frac{(-N \cdot P)^2}{2!} \right) \cdot t^2 + \left(\frac{(-N \cdot P)^n}{n!} \right) \cdot t^n)L \tag{5}$$

Since $NPt < 1$, because t = 1 year, $P \leq 1$ and N is most of times < 1, Eq. (5) becomes:

$$R = N \cdot P \cdot t \cdot L \qquad\qquad (6)$$

The risk R is the measure of a loss. For each type of loss, L_1 to L_4, the relevant risk is evaluated. The risks to be evaluated in a structure may be:

- R_1 - Risk of loss of human life;
- R_2 - Risk of loss of service to the public;
- R_3 - Risk of loss of cultural heritage;
- R_4 - Risk of loss of economic value.

The risks to be evaluated in a service may be:

- R_5 - Risk of loss of human life;
- R_6 - Risk of loss of service to the public;
- R_7 - Risk of loss of economic value.

The sum of the risk components, which may be grouped according to the source of damage and the type of damage, is named as the total risk, R.

To evaluate R, the relevant risk components (partial risks depending on the source and on the type of damage) are defined and calculated.

It is the responsibility of the national authority having jurisdiction to identify the value of a tolerable risk, R_T.

Representative values of R_T, where lightning flashes involve loss of human life or loss of social or cultural values, are reported in Table **3**.

Table 3: Typical values of tolerable risk R_T [2]

Case	R_T
Loss of human life	10^{-5}
Loss of service to the public	10^{-3}
Loss of cultural heritage	10^{-3}

The procedure to evaluate the need of protection measures considers the following steps:

- Identification of the components R_i which make up the risk;
- Computation of the identified risk components R_i;
- Computation of the total risk R;
- Identification of the tolerable risk R_T;
- Comparison of the risk R with R_T.

If $R \leq R_T$, the protection measures are not necessary. If $R > R_T$, the protection measures must be adopted in order to reduce R until $R \leq R_T$ can be achieved.

Rolling Sphere Method

The lightning strike model considered in series IEC 62305 is based on the rolling sphere method (RSM) which can be used to find the vulnerable points in a structure due to lightning flashes to ground.

The RSM is a particular case of the so-called electrogeometric model. The name "rolling sphere", originated from USA where a study dealt with this method in the early 1970s [19]. The rolling sphere radius r (striking distance or final jump distance) is correlated with the peak value of the current I. In an IEEE working group report, Estimating Lightning Performance of Transmission Lines II [20], the experimental relation is given by:

$$r = 10\, I^{0.65} \tag{7}$$

The sphere radius (20, 30, 45, and 60 m) is chosen depending on the lightning protection level, which is indentified by the risk analysis described before.

A radius of 20 m is recommended for the protection of extremely large structures or for housing explosive, flammable or valuable contents.

So far it is the best model we have to work in an international standard [15].

The evaluation of the vulnerable points over a structure is essential to design an efficient LPS against lightning flashes to ground.

Software Tools

The method for risk assessment on a structure or on a service due to lightning flashes to ground, based on IEC 62305-2, requires a great number of input data and computation.

The following tasks concerning the relevant data and the characteristics are necessary:

- For the building itself and its surroundings (Table **4**);
- For internal electrical systems, relevant incoming power line and internal electronic systems (Table **5**);

The definition of zones in the structure and their characteristics are also necessary, taking into account:

- Type of soil outside and inside the structure;
- Risk of fire;
- Existence of spatial shields;
- Existence of sensitive electronic systems;
- Type of losses.

Characteristics of defined zones are reported in Table **6**.

Being these tasks completed, mathematical computations and comparisons must be done to find the risk factors.

Table 4: Data for structure characteristics [2]

Parameter	Symbol
Dimensions (m)	(L.W.Hb)
Location factor	C_d
LPS	P_B
LPS shield	K_{S1}
Lightning flash density	N_g

Table 5: Data for internal and incoming services [2]

Parameter	Symbol
Length (m)	L_c
Aerial	-
Height (m)	-
HV/LV Transformer	C_t
Line location factor	C_d
Line environment factor	C_e
Line shielding	P_{LD}
	P_{LI}
Internal wiring precaution	K_{S3}
Equipment withstand voltage Uw	K_{S4}
SPD set	P_{SPD}
End "a" line structure dimensions (m)	(L·W·Ha)
Structure "a" location factor	C_{da}

Table 6: Data for characteristics of zones Zx considered [2]

Parameter	Symbol
Soil type	r_a
Risk of fire	r
Special hazard	h
Fire protection	r_f
Shock protection	P_A
Spatial shield	K_{S2}
Internal systems	-
Loss by touch and step voltages	L_t
Loss by physical damages	L_f
Loss by failure of internal systems	L_o

Many engineers and technicians consider the method proposed in IEC 62305-2 to be impractical, because they need: 3D reduced scale buildings; and some values of areas obtained by drawing complex shapes, which are not easy to draw. Hence, they spent a lot of time obtaining reduced scale buildings, which is a burden task and does not give enough information to find out how to protect the building properly.

Due to these facts, it is recommended to use a computer to manage all computations, to reduce errors and to avoid the burden task. To overcome these difficulties, computer tools are needed and one available tool is LPS 2008 [21], which runs over AutoCAD and is able to perform the risk assessment due to lightning flashes to the ground on a structure and on services entering in.

In Fig. **23**, a display of the LPS 2008 running over AutoCAD is shown. LPS 2008 easily extracts numeric values from objects drawn in the space model, like areas, lengths, coordinates, *etc.*After the risk assessment is completed, the user knows which level of protection must be used for the structure under study. The appropriate selection is made in the display shown in Fig. **23**. In this display, the user can define the resolution for the simulation. However, the higher in the resolution is the larger in the computation time. The dimensional information of the structure is given by the 3D model by selecting two points on the screen, to define a rectangle which involves the structure, and the highest point of the structure.

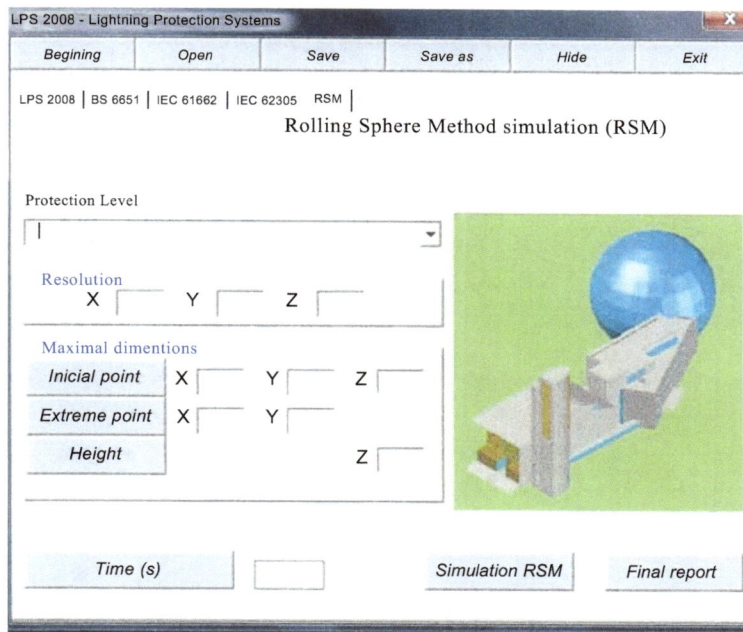

Figure 23: LPS 2008 running over AutoCAD [21].

The coordinate points are extracted to LPS 2008, then the RSM simulation is started, and during this simulation some layers and 3D objects are created automatically.

Fig. **24** shows the result of the RSM simulation on wind turbine with LPS 2008 for two different protection levels.

In Fig **24** the contour lines shown represents the boundary between the external area exposed to direct strokes and electromagnetic interference and the interior area only exposed to the electromagnetic interference.

The smallest contour line was obtained for a protection level of 98 % of effectiveness and the greater contour line was obtained for a protection level of 80 % of effectiveness. The vulnerable points to direct strokes are marked on blades and nacelle. These points should be protected.

One problem normally faced is due to the dimension of the structure, or set of structures: if the dimensions are considerable, then the computer can run out of memory during program execution, because hundreds of thousands of points will be checked.

To avoid this problem, LPS 2008 divides those dimensions in smaller parts and saves the result at the end of each computation. This new procedure allows the simulation of those structures without requiring special PC features.

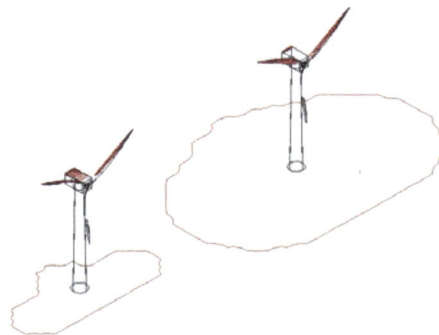

Figure 24: RSM simulation with LPS 2008 for level of protection I and IV [21].

Wind Turbine Protection

Blades are large hollow structures manufactured of composite materials, such as glass reinforced plastic. It is expected that a lightning would not strike blades made of non-conducting material, but practical experiences have clearly demonstrated that this is not the case.

Lightning does in fact strike blades without any metallic components and, whenever a lightning arc is formed inside the blade, damage is severe.

The generic problem of lightning protection of wind turbine blades is to conduct the lightning current safely from the attachment point to the hub, in such a way that the formation of a lightning arc inside the blade is avoided.

This can be achieved by diverting the lightning current from the attachment point along the surface to the blade root, using metallic conductors either fixed to the blade surface or inside the blade.

Another method is to add conducting material to the blade surface material itself, thus making the blade sufficiently conducting to carry the lightning current safely to the blade root. Variations of both these methods are used with wind turbine blades [2].

Fig. **25** shows some solutions of protection that can be used on blades against direct lightning effects.

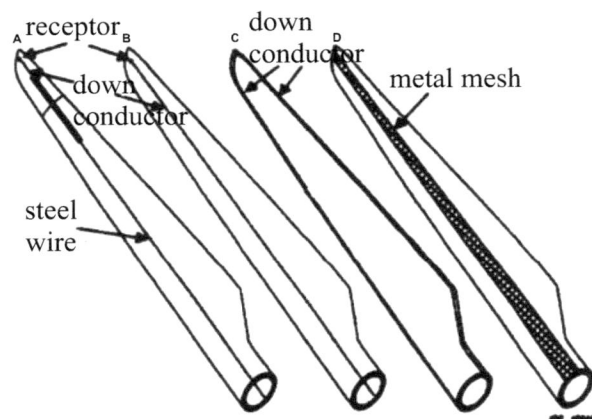

Figure 25: Lightning protection for large modern wind turbine blades [2].

However, metallic blades can also create Electromagnetic Interference problems in power converters connected to the generator.

An alternative to a lightning air termination system placed on the blade surface is to make the surface conducting by itself.

In the aircraft industry, lightning protection of glass and carbon fibre composite material for wings and surfaces exposed to lightning is achieved by adding conducting material to the outer layers, thereby reducing damage to a small area at the attachment point.

The conducting material may be metal sprayed onto the surface, metal coated fibres in the outer layers of the composite material, metal wire woven into the outer layers of the composite material, or meshes of metal placed just beneath the surface [2].

Lightning current should not pass through bearings because it usually causes damages on their surface. Bearings are not normally checked after lightning strikes, in cause of this, later faults with bearings are not linked to lightning strikes. To divert the lightning current from bearings one or more sliding contact of low impedance can be mounted between the conducting materials on blades and the nacelle, and between the nacelle and the tower. Stem discharges can be used with this purpose with the advantage that there is no physical contact, but the air gap should be minimal or the electrical impedance will be higher than that between rollers and raceways.

To disperse lightning currents flowing from a wind turbine into the earth, it is necessary to provide a suitable earth termination system to limit overvoltages that can be dangerous to both humans and equipment.

This is achievable by the provision of a low impedance earth termination system. Each wind turbine must be equipped with its own earth termination system, even if it is connected to a larger wind farm earthing system.

A wind turbine normally uses a ring electrode placed around and bonded to the foundation reinforced concrete. Horizontal electrodes may be used to connect the earthing system of one wind turbine to the next when it is within a wind farm [2].

As indicated before, lightning conductors are provided to conduct harmful currents safely into the earthing system avoiding damages due to direct effects.

Protection against indirect effects of lightning can be provided with surge protective devices (SPD), particularly at zone boundaries. An SPD or a surge arrestor limits voltages and diverts surge currents by changing its impedance above a set voltage. Normally, the device recovers its normal state after a transient has passed. The voltage at which it changes its impedance should be high enough to allow safe, normal operation, of the protected device, but low enough to avoid exceeding insulation breakdown or device failure.

Fig. **26** shows a typical electrical scheme of the external part of main substation with SPD. Shielding and bonding are also protection measures used to protect sensitive circuits against indirect lightning effects.

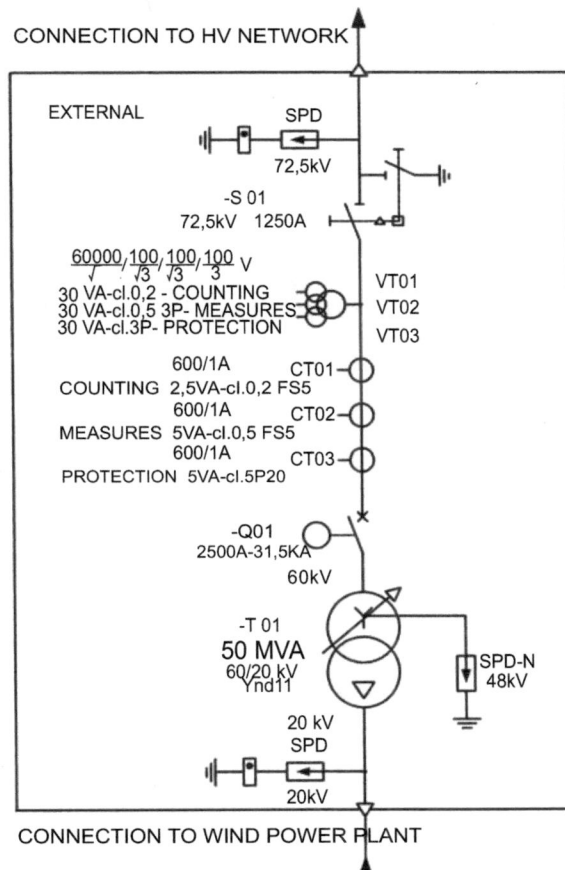

Figure 26: External part of main substation with SPD (Source: Siemens).

CONCLUSIONS

Wind power generation undergoes rapid growth and lightning incidents involving wind power systems have come to be regarded as a serious problem. Lightning protection of wind turbines presents problems that are not normally seen with other structures and will necessitate intensified discussion to improve its effectiveness. LLS can contribute with more accurate GFD maps of particular regions where investors intend to install wind farms. In order to reduce costs with protection measures against lightning it is important to perform an appropriate risk analysis, which will benefit of accurate GFD maps. Additionally the most recent standards of IEC point to the RSM as the best method to predict the vulnerable points on a structure. Finally, it has been shown that computer tools can help managing data and performing calculations and simulations, which will lead to more effective and costless LPS.

REFERENCES

[1]   Yasuda Y *et al.*, Surge analysis on wind farm when winter lightning strikes, IEEE Trans Ener Conver 2008; 23: 257-262.

[2]   IEC, Wind turbine generation system—24: Lightning protection. Tech. Rep. TR61400-24, 2002.

[3]   Rachidi F *et al.*, A review of current issues in lightning protection of new-generation wind-turbine blades. IEEE Trans Indust Electr 2008; 55: 2489-2496.

[4]   Yasuda Y *et al.*, Analysis of lightning surge propagation in wind farm. Electric Engin Japan 2008; 162: 30-38.

[5]   Lalande P. Etude dês Conditions de Foudroîment d'une Structure au Sol, Doctoral Thesis, University of Paris Sur, Septembre 1998.

[6]   Blakeslee RJ, Christian HJ, Vonnegut B. Electrical measurements over thunderstorms. J Geophys Res 1989; 94: 3135-3140.

[7]   Rakov VA, Uman MA. Lightning: Physics and Effects, Cambridge University Press, 2007.

[8]   Malan DJ. Physics of Lightning, The English Universities Press, London, 1963.

[9]   Cummins KL, Murphy MJ, Tuel JV. Lightning detection methods and meteorological applications, in: IV International Symposium on Military Meteorology, Malbork, Poland, September 25-28, 2000.

[10]   Krider EP, Noggle RC, Pifer AE, Vance DL. Lightning direction finding systems for forest fire detection. Bull Am Meteorol Soc 1980; 61: 980-986.

[11]   Lewis EA, Harvey RB, Rasmussen JE. Hyperbolic direction finding with sferics of transatlantic origin, J Geophys Res 1960; 65:1879-1905.

[12]   Lee ACL. Ground truth confirmation and theoretical limits of an experimental VLF arrival time difference lightning flash locating system. Quart J Roy Meteor Soc 1989; 115:1147-1166.

[13]   Casper PW, Bent RB. Results from the LPATS USA National lightning detection and tracking system for the 1991 lightning season, in Proc. 21st Int. Conf. Lightning Protect., Berlin, 1992.

[14]   IEC, Protection of structures against lightning—Part 1: General principles, 62305-1, 2006.

[15]   IEC, Protection of structures against lightning—Part 2: Risk management, 62305-2, 2006.

[16]   IEC, Protection of structures against lightning—Part 3: Physical damage and life hazard, 62305-3, 2006.

[17]   IEC, Protection of structures against lightning—Part 4: Electrical and electronic systems within structures, 62305-4, 2006.

[18]   Paolone M *et al.*, Models of wind-turbine main shaft bearings for the development of specific lightning protection systems, in Proc. IEEE Power Tech 2007 Conference, Lausanne, Switzerland, July 1–5, 2007.

[19]   Szedenik N, Rolling sphere – method or theory? J Electrost 2001; 51-52: 345-350.

[20]   Tehead JT *et al.*, Estimating lightning performance of transmission lines. II. Updates to analytical models. IEEE Trans Power Deliv 1993; 8: 1254-1267.

[21]   Rodrigues RB, Mendes VMF, Catalão JPS. Estimation of lightning vulnerability points on wind power plants using the rolling sphere method. J Electrost 2009; 67: 774-780.

Index

A

Analytical function of a complex variable, Chapter 3, page 66
Antenna, Chapter 3, page 72

B

Blades Chapter 5, page 106
Bearing current, Chapter 4, page 95
Bearing threshold voltage, Chapter 4, page 96
Broadband model, Chapter 4, page 80

C

Calibrated transfer impedance of the detection probe, Chapter 2, page 45
Calibration process, Chapter 2, page 57
CL-configuration filter, Chapter 2, pages 41, 57
CM chokes, Chapter 2, pages 43, 57
CM noise source impedance, Chapter 2, pages 43, 44, 47, 52, 54, 57
CM output impedance, Chapter 2, pages 48, 58
Common mode chokes, Chapter 2, pages 43, 57
Common mode inductors, Chapter 2, pages 43, 57
Common mode noise source impedance, Chapter 2, pages 43, 44, 47, 52, 54, 57
Conducted Electromagnetic Interferences, Chapter 2, pages 41, 44, 57
Conducted EMI, Chapter 2, pages 41, 44, 57
Conducted emissions, Chapter 2, pages 41, 44, 57
Coupling capacitors, Chapter 2, pages 43
Capacitive coupling, Chapter 1, page 3
Common mode coupling, Chapter 3, page 65
Common mode current, Chapter 1, pages 3, 4, 5, 9, 14, 21, 22, 25, 30, 31
Core material: Chapter 1, pages 5, 7, 8, 9, 25, 34, 37
Curie temperature, Chapter 1, pages 6, 8, 9
Cable shield, Chapter 4, page 85
Characteristic cable impedance, Chapter 4, page 87
Common mode current, Chapter 4, pages 80, 94, 95
Common mode impedance, Chapter 4, pages 84, 85, 87, 94, 97
Common mode transfer impedance, Chapter 4, page 85
Common mode voltage, Chapter 4, page 94
Conducted emission, Chapter 4, page 80
Critical cable length, Chapter 4, page 86
Control and safety systems, Chapter 5, page 108

D

Design, Chapter 1, pages 3, 5, 7, 9
Designable parameters, Chapter 1, pages 9, 10, 19, 23, 30, 33, 37
Deviation, Chapter 1, pages 23, 24, 25, 33
Differential mode current, Chapter 1, pages 9, 21, 22
DirectFET package, Chapter 3, page 62
Dielectric permeability, Chapter 4, page 87
Differential mode current, Chapter 4, pages 80, 84
Distributed model, Chapter 4, page 87
Distributed parasitic capacitances, Chapter 4, pages 84, 85, 94

Y